Nashville Native Orchids

The green glow of *Tipularia discolor* at night

Nashville Native ORCHIDS

Astonishing Science and Mysterious Folklore

SORAYA CATES PARR

VANDERBILT UNIVERSITY PRESS
Nashville, Tennessee

Published 2024 by Vanderbilt University Press

First printing 2024

Library of Congress Cataloging-in-Publication Data

Names: Parr, Soraya Cates, 1959- author.
Title: Nashville native orchids : astonishing science and mysterious folklore / Soraya Cates Parr
Description: Nashville, Tennessee : Vanderbilt University Press, 2024 | Includes bibliographical references
Identifiers: LCCN 2024006701 | ISBN 9780826506580 (paperback) | ISBN 9780826506597 (epub) | ISBN 9780826506603 (pdf)
Subjects: LCSH: Orchids--Tennessee--Nashville. | Endemic plants--Tennessee--Nashville.
Classification: LCC SB409.5.U6 P37 2024 | DDC 635.9/3472--dc23/eng/20240329
LC record available at https://lccn.loc.gov/2024006701

Cover images: Front, *Spiranthes cernua*; back, *Spiranthes lacera* var. *gracilis*. Photographs by Soraya Cates Parr

Dedicated to the memory of C.C. Cates, Jeannette H. Cates, Almaz K. Homra, and Linda S. Shumate.

Their unique and creative lifestyles instilled the love of nature into the consciousness of a child.

39
53
67
81
95
109

CONTENTS

FOREWORD

IN EVERY GENRE OF NATURALIST ENCHIRIDION, the most recent works become repetitive of their predecessors. Field guides and reference books on a particular topic tend to contain similar content and just scratch the surface of the natural history of the flora or fauna. Occasionally, however, an inspired writer presents a more detailed look into the life of a plant or animal and gives the reader a glimpse into the true nature of the species. *Nashville Native Orchids* by Soraya Cates Parr is an inspired, detailed piece of literature.

Nashville Native Orchids presents a comprehensive look into the secret lives of common orchid species. Most of the plants are easily seen in winter with their green leaves since summer and early fall flowers are often overlooked. Soraya easily identifies orchids in all seasons and knows which woodland categories and tree species they prefer. The book includes information on interactions with soil organisms, animals as seed dispersers, all steps of seed production, and niches within habitats and geographical areas.

Each chapter includes vibrant, unique photographs of the plant's life cycle. The chapters tie in with cultural connections from various time periods. In addition, Parr's nocturnal research and photos show the associations with nighttime creatures; most research by other botanists is conducted more during the light of day. The reader becomes absorbed in the story of each orchid and the interactions with other components of nature. This is not simply a book of rote facts and monotonous information. It is a book by a scientist who is an artist at heart.

I met Soraya several years ago in a naturalist setting and quickly became fascinated with her vast knowledge on numerous topics, especially native orchids. Her passion for these plants goes well beyond that of researcher or orchid enthusiast. She has a connection to them in a way that makes the drive for their conservation a daily life pursuit. She can be found at any hour of day or night in the woods looking at their pollination or seed-dispersal vectors. She has a lifetime love for orchids, especially those native to Tennessee.

My background as the former Tennessee State Parks biologist led me to working in many of the Metro and State natural areas and parks. As the director of Beaman and Bells Bend Parks, I have managed many acres of woodland and habitat types. Never have I met a more principled activist for conservation and the study of wild lands. There is a great need for this book to educate other park naturalists to plants of which they have little knowledge.

Nashville and similar cities need available literature on irreplaceable native flora. Excessive growth and unbridled development destroy unique plant communities and rare species. Most Nashvillians do not recognize the treasures in their own backyards or local parks. The bulldozer does not distinguish the life being destroyed as it flattens all in its path. Natural and cultural resources have no influence on the mass destruction caused by a growing city. A specialized book such as this about little-known plants can enlighten the public and lead others to preserve individual plants or entire stretches of the forest.

Naturalists at a nature center armed with the knowledge of such a book can educate large numbers of citizens and encourage them to join in conservation efforts. Schoolchildren aware of native orchids from an early age will want to protect them.

An indication of an advanced stage of moral maturity is to care for the well-being of others instead of solely focusing on the self. A true conservationist recognizes the intrinsic value of nature and strives to help species that cannot help themselves. Plants or animals do not have to benefit humans to have worth. They have importance for being living creations on our planet. *Nashville Native Orchids* not only opens closed eyes to the wonders right outside our doorways, but it also encourages the public to take the next step toward a maturity focusing on the conservation of precious, invaluable plants and animals.

LinnAnn Welch, MS
Director, Bells Bend Outdoor Center
Nashville Metro Parks and Recreation

INTRODUCTION

The Hidden Treasures of Nashville

IF YOU OPEN A ROAD ATLAS to the state of Tennessee, and look in the middle of the map, you will find the city of Nashville. This growing metropolis is in Davidson County on the Cumberland River. The river snakes its way around the geophysical regions of the Highland Rim, with Nashville positioned in the Central Basin. This elliptical bowl is surrounded by steep hills and deep valleys. Antiquated history books describe this basin as an area where a large part of the city of Nashville was built over an extensive stone-grave cemetery.[1]

Nashville has an abundance of beautiful woodlands and wildlife in surprisingly close proximity to residential homes. Deciduous trees are lush in the spring and colorful in the autumn. The four seasons bring a variety of plant life. Wildflowers of every color can be found throughout the year. Blackberries and persimmons provide fruit, while ferns and mosses delight with their feathery foliage.

The official animal representing the state of Tennessee is the raccoon, *Procyon lotor*. It roams freely along with the deer, opossum, squirrel, and woodchuck. Many species of birds soar in the sky, while aquatic birds catch their meals of fish in small ponds.

FIGURE 0.1. A luna moth (*Actias luna*) visits a *Tipularia discolor* at night. All photographs by Soraya Cates Parr.

Music City, USA, is considered the home of country music. Fans from all over the world embark to this southern empire. It is not unusual to see a pair of custom cowboy boots sashaying their way to the various music venues that line the downtown streets.

History buffs come to explore sites such as Andrew Jackson's Hermitage and the Ryman Auditorium. Sports enthusiasts enjoy the roaring crowd at a Titans football game or experience the fast pace of the Predators hockey team in action. Hungry travelers can order up good ole' Southern cooking such as fried crunchy-crust chicken, biscuits with sawmill gravy, and beans, with a little coleslaw on the side.

Along the scenic Natchez Trace Parkway a traveler will see outcrops of limestone jutting from their blue-gray strata. You can imagine hearing these ancient stones whispering tales of the Civil War battles fought throughout this area of Middle Tennessee.

Not very far away from the sounds of laughter and guitars, the crowded sports stadiums, busy restaurants, and historic memorials, Nashville holds an unseen treasure. The real jewels are the orchids that flourish here, the Nashville native orchids.

Middle Tennessee has an abundance of native orchids; however, this guide will describe the commonly encountered orchids within the city limits of Nashville. You can locate these hidden treasures in greenways, parks, older cemeteries, and in your own suburban backyard.

This orchid lineup is impressive: *Galearis spectabilis*, *Goodyera pubescens*, *Liparis liliifolia*, the *Spiranthes cernua* complex, *Tipularia discolor* (fig. 0.1), and the occasional *Cypripedium parviflorum* var. *pubescens*.

1

Basic Orchid Science

Monocotyledon and Dicotyledon

The orchid family belongs to a group of flowering plants known as monocotyledons, or monocots. Monocot seedlings have a single seed leaf, or cotyledon, upon germination. Monocots have parallel veins with the central vein extending the length of the leaf blade. The flora parts are generally in threes or multiples of three.[1] Other examples of this group are grasses, irises, lilies, and sedges.

Dicotyledons, or dicots, have two seed leaves. Dicots commonly have branched veins that form a network. The flora parts are in fives or multiples of five.[2] Examples are daisies, roses, and significant food-producing plants such as beans and cabbages.

The Divine Design: Photosynthesis in Orchids

Photosynthesis is the process by which plants can manufacture their own food by utilizing light energy. It occurs in two stages: the light reaction, which requires the absorption of light, and the dark reaction, or Calvin cycle, which does not require light energy.[3]

The process of the light reaction begins with the raw materials of carbon dioxide and water. Carbon dioxide is absorbed through the stomata of the leaves, while water is absorbed through the roots in the soil. The rigid cell walls within the orchid leaves contain photosynthetic organelles, the chloroplasts. Chloroplasts contain the principal green pigment, chlorophyll. Chlorophyll mediates light energy from the sun and transfers it into chemical energy. This promotes a chemical reaction that splits the water molecule into hydrogen and oxygen. Oxygen is liberated through the stomata into the atmosphere, while some is used by the plant for the process of respiration. The hydrogen is incorporated into the production of simple sugars during the second stage, the dark reaction or Calvin cycle. During this process, carbon dioxide is reduced into sugars to be used by the orchid. Simply stated, photosynthesis is a means for plants to make fuel for their growth and development.

The Unique Characteristics of Orchids

The Orchidaceae are one of the largest and most diverse families in the plant kingdom. At the root level, corms, pseudobulbs, rhizomes, and tubers function as storage organs for carbohydrates and water. The leaves and flowers develop by utilizing food stored in these underground components.

The orchid has three sepals and three petals.

The sepals typically form the outer cover of the orchid flower bud. As the bud opens, the sepals may differentiate in shape and color from the petals. A large and impressive dorsal sepal can develop in some species.

Two petals are generally alike, while the third petal is modified into a highly specialized organ called the labellum or lip. Variations of the labellum show great diversity as they can be fringed, hooked, lobed, pouch-like, notched, or bizarrely shaped. They are sometimes modified to include a spur at the base to store nectar.

The orchid contains a specialized column in the central portion of the flower. The fused stamen (male part) and style (female part) are united into a single structure. Columns vary in shape and size. For example, the column of *Liparis liliifolia* has an insect-like profile.

The pollen produced by the orchid are spherical, wax-like masses. They are contained in pollen packets called pollinia (sing., pollinium). Frequently, the pollinia are attached to a sticky pad called a viscidium.

Orchids have an inferior ovary. Within this chambered organ are abundant dust-like seeds, each with an embryo surrounded by a thin, dry seed coat.

These characteristic features demonstrate that orchids are some of the most highly evolved plants in the world.

The Role of Fungi in Orchid Survival

Fungi play an important role in the orchid's life cycle. The fungi are associated with the roots of higher plants such as forest trees.[4] They flourish in the decayed leaf litter and soil. Since the orchid embryo contains

FIGURE 1.1. Microscopic view of the fungal invasion of a cluster of *Goodyera pubescens* seeds.

limited or no endosperm, it must rely on a symbiotic fungus during seed germination for an adjunct source of nutrients. This unique relationship between orchids and fungi is called a mycorrhizal symbiosis association.

The process begins with the invasion of strands of fungal hyphae entering the cells of the orchid roots or surrounding the base corms (fig. 1.1).[5] Inside the cells, the fungi form structures called pelotons that can be used by the orchid to support germination. An equilibrium is established as the orchid obtains the needed nutrients for germination and development, while the fungal component has a suitable location for growth and reproduction.[6] Both fungi and orchid benefit from the symbiotic union.

Mycorrhizal fungi develop extensive networks of string-like hyphae within the soil. Sizable networks connect numerous plant communities and help to supply nutrients to host plants. Modern genetic tools have revealed that fungi and plants use these pathways in communication processes. They signal messages between the fungi and the orchid cells.[7] However, mutual symbiotic associations between mycorrhizal fungi and orchids do not always occur. Conversely, certain fungi may become pathogenic parasites. They are capable of secreting harmful enzymes that damage the embryo if conditions such as appropriate levels of available nutrients in the soil, humidity, light, and temperature are not met.[8]

This symbiotic union was described by the French botanist Noël Bernard (1874–1911). While walking in the Fontainebleau forest, he found a dried inflorescence of an achlorophyllous orchid.[9] This presented a new idea to him on orchid seed germination. His theory stated that orchid seeds do not germinate in the absence of certain fungi. He presented his ideas to the French Academy of Sciences with mixed results of support and skepticism from his peers. With the cultures of fungi Bernard cultivated, he successfully demonstrated the necessity of fungi in the germination of orchids.[10]

Pollination Mechanisms of Orchids

The evolutionary mechanisms involved in orchid pollination are both complex and diverse. Floral signals and plant deception are two methods orchids use to help ensure cross-pollination.

Floral signals include color, inflorescence shape, and fragrance. An example of an orchid that uses both color and shape is *Cypripedium parviflorum* var. *pubescens*. Potential pollinators are attracted to the bright color and large pouch-like structure of the labellum from far distances.

The shape of the *Tipularia discolor* inflorescence provides another unique floral signal. The stately inflorescence produces an illusion of crane flies hovering around a central axis. This form of deceit attracts insects for mating purposes. Landing on the flowers, they inadvertently gather pollinia and transfer it to an adjacent flower.

Orchids deceive potential pollinators by their fragrance. They produce an odor that contains a pheromone, a chemical messenger similar to that produced by a female insect. When pheromone-like odors are released from an orchid, male pollinators will try to copulate with the flower. During the pseudocopulation, pollinia become attached to the eyes, head, or abdomen and are transferred to another orchid. Often, pollinia are deposited in the stigmatic surface, aiding effective pollination.

Orchid species are either self-compatible or self-incompatible. Orchids that are pollinated successfully without the aid of insects or other types of pollinating agents are termed autogamous.[11] These self-compatible orchids are fertilized by their own pollinia and produce seed. Weather conditions, such as rain and wind, facilitate fertilization by bringing pollinia into contact with the stigma. However, self-fertilization may diminish genetic variations, which can reduce plant hardiness and vigor over time.[12]

Self-incompatibility mechanisms prevent asexual fertilization. An example of this method would be a modification in the design and structure of the column that would prevent the transfer of pollinia grains to the stigma; thus, pollination would not occur.

Orchid flowers have distinct color, fragrance, and structural characteristics that are a result of adaptation to the behavior and traits of the pollinator.[13] A number of insects are attracted by the colors and fragrances, primarily bees, flies, moths, and wasps (fig. 1.2). Flowers pollinated by birds tend to be brightly colored. Nectar-producing flowers are adapted to butterfly and moth pollination.

Pollination mechanisms were studied by Charles Darwin. He became fascinated by the orchids that grew along the chalk hills near his home in Kent, England. His work *The Various Contrivances by Which Orchids Are Fertilized by Insects* was published in 1862. His research involved the study of the modifications of the orchid flower to aid effective cross-pollination.[14] He considered the adaptations of orchid flowers as being the paramount example of evolution through natural selection.[15]

FIGURE 1.2. A *Bombus* species pollinates *Spiranthes cernua*.

FIGURE 2.1. A terrestrial *Tipularia discolor*.

2

Terrestrial and Tropical Orchids

What Are the Differences?

Terrestrial Orchids

Terrestrial orchids grow on or near the soil surface (fig. 2.1). They flourish in leaf litter, decaying humus, and in a variety of soils. They provide shelter, nectar rewards, and unique ecological communities for the tiny arthropods that live within a few millimeters of each other on the orchid. Ants, bees, wasps, and flies seek nectar rewards. Spiders spin delicate webs to catch their prey. Microscopic insects aerate the soil around the corms and roots. When humans destroy the habitat of a single orchid, multitudes of other life-forms perish.

Tropical Orchids

Tropical orchids are exotic and inspirational (fig. 2.2). Throughout the centuries, they have been collected, hybridized, painted, photographed, and cultivated in glass greenhouses. Large concentrations of tropical

FIGURE 2.2. A tropical *Phalaenopsis*.

species are found throughout the world, including Central and South America, China, India, and Southeast Asia.

Tropical orchids are primarily epiphytic. Specialized aerial roots do not parasitize; they support the orchid by anchoring to trees, rocks, and other nearby plants. Adapted for the absorption of moisture, tropical orchid roots are hydrated by rain and the surrounding humidity. Nutrition is obtained from plant litter and decaying insect matter.

Years ago, tropical orchids were collected from the wild and hybridized to develop interesting colors and hardy characteristics. Today, large-scale propagation is carried out in high-tech laboratories.

3

Orchid Myths, Rituals, Traditions, and Uses

THROUGHOUT HISTORY, THE ORCHID has been revered as a magical and sacred gift from the celestial gods. Orchids are associated with herbal cures, love charms, and spiritual themes. Their brilliant colors, fragrances, and fascinating growth habits excite the senses of people from every culture.

Early alchemists hypothesized that the shape of a plant would provide a diagnostic key to alleviating disease or pain in the human body. Greek physicians interpreted the shape of orchid twin tubers to resemble the male anatomy. Thus, the tubers were used as aphrodisiacs. Orchids are listed in antiquated medical manuals for use in fertility potions and to treat nerve inflammation and insomnia.

The Orchid in Ancient Mesoamerica

The orchid was considered a dignified floral symbol in ancient Mesoamerica. It is found in primeval artwork and sculpture as far back as the Teotihuacan culture, approximately 200 to 900 AD. We know very little about the people who lived during the rise of Teotihuacan. Archaeological studies show that the Aztec civilization believed it was a place of mystery and religious practices.[1]

The ceremonial city of Teotihuacan was located in the central valley of Mexico. Significant religious Mesoamerican structures, including the Pyramid of the Sun and the Pyramid of the Moon, were built in the pre-Columbian period.[2] Sacred rituals performed at these sites honored the deities of nature, including the moon, stars, and sun. These practices guaranteed the fertility and prosperity of the people. Although stories of horrific bloodletting and human sacrifice were the common exchange, the rituals described in those stories were not the only type of ceremony performed. Tropical flowers such as bromeliads and orchids were venerated as sacrificial offerings. Today, the remnant structures of Teotihuacan are the reminders of a lost civilization.

The vanilla orchid was very important to the early Mesoamericans. This orchid produced a seed pod that was pulverized and mixed with chocolate.[3] Chocolate contains alkaloids that are stimulating to the central nervous system. This flavor-filled treat had the added advantage of enhancing physical strength and endurance.

Orchids play an important part in many rites of passage. In Mexico, and especially in Oaxaca, orchids along with marigolds are displayed in the traditional celebration of Día de los Muertos. On the second of November, souls of the departed return to visit their loved ones. Families enjoy a day decorating family graves. Picnics are held late into the night.[4] *Ofrendas*, or altars, are constructed in homes with great celebration. Favorite flowers, fruits, and foods are displayed along with pictures of the deceased. Many times, a single white orchid, symbolizing purity, is placed upon the *ofrenda*.

In Mexico City, December 12 is a holy feast day honoring the Saint of Mexico, Our Lady of Guadalupe. Visitors from all over the world travel to join in the celebration. They bring offerings of orchids, roses, and lilies to show their reverence.

An interesting legend circulates about this event. In 1531, Our Lady of Guadalupe appeared to a craftsman named Juan Diego. She asked him to instruct the bishop to build a shrine in her honor. She wished to bless and protect the impoverished native people of the land. When the bishop insisted on visual proof, Juan brought in his simple shawl, a *tilma*, to show him. It was filled with roses he had picked while conversing with the Virgin. Since it was winter, roses were out of season. An image of the Virgin, with tears in her eyes, appeared on the *tilma*. The appearance of the image, along with the roses, convinced the bishop to appoint workers to build the shrine. Today the *tilma* is framed and hung over the altar for all to see.[5]

The Orchid throughout the World

Spanning the globe from ancient Mexico to old Ceylon, the Wesak orchid, *Dendrobium maccarthiae*, is associated with the birth of Buddha.[6] When the moon becomes full, the Wesak Festival of Lanterns is celebrated with orchids, fruits, and hundreds of lighted lanterns.

In the Philippines, orchids from the *Dendrobium* species have extremely tough and durable fibers for making baskets, clothing items, purses, and sandals. Sales from these products may secure a financial future for a family.[7]

In India, Asia, and the Mediterranean, orchid tubers are used as flavorings for ice cream. Salep is the name of the starchy flour made from orchid tubers. Local shepherds, who gathered the tubers for sale, have confirmed that populations are decreasing.[8] Today, new conservation measures have discouraged the use of these orchid tubers. Aficionados of salep, however, declare that no other alternative flavor can compare.

In China, orchids have been cultivated for thousands of years. The flowers are depicted in various art forms, from sculpture to wall

FIGURE 3.1. The terrestrial *Tipularia discolor* in the wild.

paintings. Several species of orchid tubers are still used in traditional Chinese herbal medicines today.

In some countries, orchids are regarded as plants that can understand the emotions of the human heart. Recently, the author had a conversation with a young Costa Rican who explained that his mother had instructed him not to water orchids when he was angry or sad.

His collection of orchids perished after he watered them following a tragic personal event. He believed this validated his mother's advice on orchid-watering techniques.

The Native Orchid in Early America

In North America, terrestrial orchids were used by early Native Americans.[9] Medicines made from these plants ranged from treating skin diseases to dispelling spiritual crises. The early colonists were instructed by friendly neighboring tribes on the medicinal benefits of these native plants. Medical information was passed down verbally from one generation to the next.

Native orchids were used to treat skin diseases, nervous conditions, and snake bites.[10] Corms, tubers, leaves, and flowers were pulverized, boiled, and taken as a syrup, or burned and inhaled. Poultices were prepared and left in the sun for natural fermentation to occur. This slurry was placed on the afflicted part of the body and covered with wool to penetrate the skin. Spiritual illnesses were often treated with the same type of poultice placed over the heart or the top of the head.

Today, the orchid has remained a symbol of hope, nobility, and beauty (fig. 3.1).

The Orchid Craze of the Victorian Age

The reign of Queen Victoria, from 1837 to 1901, brought about great cultural changes in Britain and the United States. This colorful era created a fondness for the unusual in art, music, and fashion.

In Britain, Old Christmas Day was celebrated on January 6, Twelfth Night.[11] This was a time of great celebration with dancing and pantomime featuring the Harlequin character. Smorgasbords of foods such as stuffed peacock were served along with the traditional boar's head. Cards and gifts were bestowed on family and friends. In the firelight, the game of Snapdragon was played. Each person had to pick a raisin from a tray of flaming liqueur without their fingers getting burned.

In America, holidays were celebrated with zest. The harvest season ushered in lavish Halloween parties with costume and dance. Mysterious games of fortune were played with fruits of the harvest: apples, cabbages, and nuts. As the last candle sputtered out in the jack-o-lantern, the party host would narrate chilling ghost stories.

With this extravagance came a desire for unusual plants, especially among the affluent. Orchids were considered the flowers of royalty (fig. 3.2). As the lure of wild, rare orchids began to accelerate, extensive travel plans were scheduled throughout the world for collection purposes. Plant collectors would be paid to search through uninhabitable terrain to find these floral gems. They fought sweltering climates and vectors of tropical diseases. Encounters with the regional people could be hazardous, as these collectors were viewed as threatening intruders. A vast number of varieties of unusual plants were collected and sent back home to grow in modern glass greenhouses.

There is an interesting story about how the first orchids became popular in the Victorian era. In 1818, an orchid aficionado, William Swainson (1789–1855), was collecting plants in Rio de Janeiro. He unknowingly used the dried pseudobulbs of orchids as a packing material for the collected plants.[12] He shipped some to his friend William Cattley (1788–1835) and some to the Glasgow Botanical Gardens. Cattley successfully flowered some of the dried pseudobulbs. He was astonished by the magnificent flowers that appeared from them.[13] Later, in 1824, Cattley was honored by the botanist John Lindley (1799–1865), who named the genus *Cattleya*.

Cattleya orchids range in a variety of colors and sizes. With their frilly, large flowers, they are the familiar orchids used in corsages (see fig. 3.3).

FIGURE 3.2. Tropical orchids like this *Phalaenopsis* were favored during the Victorian era.

FIGURE 3.3. An example of a tropical *Cattleya* orchid.

FIGURE 4.1. A *Bombus* bee forages on bear corn (*Conopholis americana*).

4

Climate Change, Sad Exchange

THE WORDS *CLIMATE CHANGE* and *global warming* often invoke heated debates. The varied results of these factors are recognized by scientists as having an adverse impact on the health of the environment as well as negative consequences on the human body.[1] Indicators such as long-term temperature tracking can provide the evidence needed to understand existing climate change.[2]

Climate change and *global warming* are used interchangeably; nevertheless, there is a difference. Climate change refers to any long-lasting significant measurement changes in climate, whereas global warming is the average increase in the temperature of the atmosphere near the Earth's surface.[3] In this guide, we will discuss the effects of global warming in detail.

An Explanation of Global Warming

Our planet receives wavelengths of solar radiation from the sun. As the radiation passes through the atmosphere, most of it is absorbed by the surface of the Earth. On receiving this sunlight radiation, the Earth emits long, infrared wavelengths back into the atmosphere. Atmospheric gases such as carbon dioxide, oxygen, ozone, and water vapor are selective absorbers of these wavelengths. They redirect some of the infrared waves back into the atmosphere where it becomes trapped. This functions much like a glass greenhouse that forms an insulating barrier around the Earth, resulting in the accumulation of heat. The additional heat causes higher temperatures in the atmosphere and in the oceans. This phenomenon is called the *greenhouse effect*.

Greenhouse gases are a result of increased human activities, such as the burning of fossil fuels to power vehicles. Examples of greenhouse gases are carbon dioxide, methane, nitrous oxide, and fluorinated gases called chlorofluorocarbons (CFCs).

Since the Industrial Revolution, pollution has caused concentrations of carbon dioxide and other gases to rise and remain in the atmosphere for long periods of time. Past emissions are affecting our planet today, while future ones will continue to increase the levels of detrimental gases.[4]

Specific Changes in Native Orchids Due to Global Warming

On the environmental scale, distant glacier melts due to global warming are causing modifications in ecosystems all over the world. Native orchids are especially vulnerable to the effects of environmental change. Some of these changes include the introduction of invasive pests and pathogens, destruction of natural habitats from rapid urbanization, disruptions in mycorrhizal associations, habitat losses, and alterations in forest compositions. These detrimental effects result in a disturbed ecosystem, which produces changes to native orchid communities.[5]

It is interesting to note that this occurrence has been documented in other parts of the world for native orchid populations, notably in Britain, Europe, and the Middle East.

A research study led by Professor Michael Hutchings at the University of Sussex has shown that with increases in temperatures, pollinators may not be synchronized with the anthesis of orchids. Therefore, successful pollination cannot occur. Professor Hutchings explains that with any small rise in spring temperature, female bees tend to achieve their peak flying time two weeks or more before the orchids reach their definitive flowering date.[6]

Specific Changes in Native Orchids Due to Rapid Urbanization

In the environmental and institutional context, conditions from rapid, poorly planned construction and sudden land-use changes have an impact that disturbs the survival of various orchid populations. Infrastructural changes in land-use from wooded acreage to a sprawling apartment complex result in abrupt land loss and inadequate drainage with runoff. These detrimental effects have caused orchid communities to be displaced as they are swept away into roads and sewage drains.

To reduce global warming, it will require urgent international cooperation, as well as the development of community-centered projects.[7] This collaboration would be beneficial in several ways: by supporting animal and plant life, protecting food and water supplies, reducing disruptions in aquatic life, and lessening the burden of adverse human health issues. If we educate ourselves about climate change, the knowledge will guide us to take important steps in conserving our native fauna and flora for years to come.

The Significance of Citizen Science Monitoring Programs

Temperature and light are crucial factors in determining the timing of the arrival of certain species of pollinators and the flowering of their preferred nectar plants. Given the important function of pollinators in the ecosystem, more citizen monitoring programs are needed. Chronicling observations may assist scientists in detecting future changes within insect pollinator communities. Here is an example of a citizen monitoring program initiated by the author:

THE EMERGENCE OF *BOMBUS* POLLINATORS

The emergence of *Bombus* queens and the maturation of the flowers of *Galearis spectabilis* were noted to be in synchronization spanning the years from 2015 to 2017. During that time period, it was noted that the intensity of fragrance was at its height concurrently with the emergence of the bees.

Between the years 2018 and 2022, the author noted the *Bombus* pollinators emerged an average of seventeen days earlier than the maturation of the flowers. Flower maturity was identified by the intensity of the floral fragrance and the heightened degree of color development in the flower. The bees foraged within a few meters in the vicinity of the orchids but preferred the highly fragrant achlorophyllous Bear Corn (*Conopholis americana*; see fig. 4.1).

After six days, the intensity of the scent was pronounced, yet the foraging *Bombus* bee population had dwindled in the orchid's vicinity.

Small bees of the Halictidae family were observed landing on the *Galearis* flowers. However, later in the season, the flowers did not produce seed pods.

As you can see by this monitoring example, this valuable information can provide a window into what is occurring in your community. Anyone can assist in a citizen science program. The main goal is to get involved and learn what you can do to help your local environment.

FIGURE 5.1. A colorful *Theridion frondeum* on a *Galearis spectabilis* bract.

5

The Ecology of Native Orchids and Spiders

WITH THE ADVENT OF GLOBAL WARMING, the word *ecology* has become a universal term in the media. Ecology is the study of the relationships between living things and their environment. Ecological studies continue to address the complex levels of biological organization in our world today.[1] In simpler terms, everything from scientific principles to political actions have consequences for the environmental life of an organism, including the activities of *Homo sapiens*. By becoming aware of our local ecology, we can recognize how our actions affect the world we live in.

Native Orchid and Spider Relationships

Throughout the world, an essential part of the orchid ecosphere is a group of arthropods in the Arachnida class, the spiders. Spiders have one of the most prolific alliances with terrestrial orchids. Orchids provide spiders with transitory to permanent habitats for capturing prey,

FIGURE 5.2. A *Verrucosa arenata* is finding sanctuary on a *Tipularia discolor* orchid.

FIGURE 5.3. A translucent green *Mangora maculata* is camouflaged next to the glowing green column of *Tipularia discolor*.

protection from weather conditions, and sites for reproduction. In turn, spiders aid in protecting orchid tissues from invading parasites.

The deciduous woodlands that surround Nashville contain sizable populations of orchid and spider species. Here are some examples:

Ghost spiders are active members of the Anyphaenidae family. As the name implies, they are usually pale in color.[2] One particular spider in this family has an interesting relationship with *Tipularia discolor*. *Wulfila saltabunda* is a semi-transparent spider that uses strong silk strands to bind the orchid leaf together into an enclosure. Within this fold, a location is provided for their protection and reproduction. The spider detects vibrations by monitoring silk strands in the fold or on the leaf surface. They attack prey that could harm leaf epidermal tissues.

The lined orbweavers *Mangora gibberosa* and *Mangora maculata* are in the Araneidae family.[3] Both species can be frequently observed on *Tipularia discolor* orchids. These translucent green spiders are decorated with shaded lines and dots. They build distinctive webs with a dense stabilimentum circular inner web surrounded by a more delicate outer web. A line of silk runs from the web to the buds and open flowers of *Tipularia*. These clever spiders wait on the column, petals, or sepals (fig. 5.3). The floral parts provide camouflaged areas that conceal the spider's location. When an insect lands on the flower, the spider pulls or shakes its silken threads. This encourages the prey to navigate closer for a swift ambush.

Another spider in the Araneidae family is *Verrucosa arenata*.[4] This arrowhead orbweaver has a preference for the orchids *Goodyera pubescens* and *Tipularia discolor*. Both orchids bloom in the mid- to late summer when *Verrucosa* are actively seeking a temporary summer habitat. Arrowhead females sport a white or yellow triangular abdomen (see fig. 5.2).

The manner in which this spider finds food is noteworthy. As an insect lands on the strong web, the female will bounce on its silken threads to locate the position of the prey. By pulling on the threads, it will reel in the scrumptious target.

Theridion frondeum belongs to one of the largest family of spiders, the Theridiidae.[5] They are commonly named the cobweb builders. This bright yellow and black spider is frequently found on the orchid *Galearis spectabilis* (fig. 5.1). The dorsal hood and nectar column provide the spider with a temporary habitat while it waits for prey.

An interesting fact about this group of cobweb builders is that they have the rare quality of social behavior. Such behavior is usually found in ants, bees, and termites. The spider's egg sacs can be found on the underside of the *Galearis* leaves. The mother spends time with the juveniles beyond the egg sac stage and demonstrates maternal instincts.[6]

Jumping spiders are in the Salticidae family.[7] They are active during bright sunlight. *Phidippus audax*, *Phidippus putnami*, and *Phidippus richmani* can be found on the pedicels and leaves of *Liparis liliifolia*, members of the *Spiranthes* genus, and *Tipularia discolor*. They make small tent-like shelters to reside in during the evening. With four pairs of eyes, they spring into action to catch their prey (fig. 5.4). They are naturally inquisitive about their surroundings. If you are taking photographs of them, they may even jump on your camera to get a better look!

Diverse Attitudes toward Spiders

The contemporary opinion on the subject of spiders is bleak. Most people cringe at the mere mention of the word *spider*. They have an arsenal of sprays handy just in case one is creeping onto the front porch! However, attitudes on the important function of spiders in gardening and native plant communities are changing. Spiders are recognized as performing a significant role as natural control agents. They contribute to the forest ecosystem by stabilizing and regulating insect populations.[8] Without spiders to protect native orchids, damage may result to the entire plant. Opportunities exist for further research on the relationship between spiders and native plant life.

FIGURE 5.4. A majestic male *Phidippus putnami* stakes his claim on a *Liparis liliifolia* leaf.

Simple Key to the Nashville Native Orchid Genera

How to Use This Key

Keys for identification can be very confusing. Do not be intimidated about using a genera key. It is a simple process of elimination.

First, determine if the orchid has green leaves while it is blooming. If yes or no, go to the number 2 choices. Next, the key asks if the orchid has variegated or patterned foliage; if it does, you will see that it fits the orchid *Goodyera pubescens*. If it does not, you proceed to the number 3 choices. Next, determine if the orchid has a showy or pouch-shaped labellum. Continue this process until you have eliminated or arrived at the key choices.

For example: *Galearis spectabilis* has green foliage at the time of blooming (proceed to choice 2); no variegated foliage (proceed to choice 3); no twisted inflorescence (proceed to choice 4); no asymmetrical arrangement (proceed to choice 5); a large labellum and no slipper-like pouch (proceed to choice 6); a large labellum and dominant hood at apex. So the logical choice is 7.

You will discover that different guidebooks have a large variety of keys available. Keys in publications can differ widely. You will become more familiar with using keys by looking at several field guides, obtainable from your public library or bookstore.

1	Orchids with green foliage (leaves) at time of flowering	2
	Orchids without green foliage (leaves) at time of flowering or foliage reduced to scales	2

2	Orchids with variegated or patterned foliage	***Goodyera pubescens***
	Orchids without variegated or patterned foliage	3

3	Flowers in a spirally twisted inflorescence	***Spiranthes cernua***
	Flowers not in a spirally twisted inflorescence	4

4	Flowers asymmetrically arranged	***Tipularia discolor***
	Flowers not asymmetrically arranged	5

5	Flowers with a large labellum and a slipper-like pouch	***Cypripedium* spp.**
	Flowers with a large labellum and no slipper-like pouch	6

6	Flowers with a large labellum and spidery inflorescence	***Liparis liliifolia***
	Flowers with a large labellum and no spidery inflorescence	7

7	Flowers with dominant hood at apex	***Galearis spectabilis***

THE ORCHIDS

FIGURE 6.1. *Galearis spectabilis* in full bloom.

6

Galearis spectabilis

A Showy Spectacle Living in Secrecy

THE TINY ORCHIDS OF SPRINGTIME, *Galearis spectabilis* (L.) Rafinesque, are located within the partially shaded woodlands that surround Nashville.[1] With a pink helmet on top and a wavy, white labellum underneath, they are considered the "movie stars" of the native orchid world (fig. 6.1).

Common names for this perennial are the Purple-Hooded Orchid and Showy Orchid. The word *galearis* is derived from Latin meaning helmet, referring to the helmet-shaped hood at the apex of the flower. The word *spectabilis*, also from Latin, means showy or spectacular. In archaic manuals, this orchid had the taxonomic classification of *Orchis spectabilis*.

This North American species extends from Quebec to Ontario and into the Great Lakes region. Traveling southward, it extends through the eastern half of the United States where it sweeps across South Dakota and Kansas, down to Oklahoma, and into the southern states. It is rarely found in Florida or Louisiana.

The Showy Orchid prefers rich soils in moist areas near streams or ponds. It flourishes in calcareous sites and is found in older flood plains. These orchids occur in communities of three to ten plants. Isolated plants may occur within a few meters of the parent colony.

HELPFUL HINT

When searching for this orchid, look for the trees of sugar maple (*Acer saccharum*), shagbark hickory (*Carya ovata*), American beech (*Fagus grandifolia*), sweetgum (*Liquidambar styraciflua*), tulip poplar (*Liriodendron tulipifera*), and oaks (*Quercus* spp.). The author has noticed a marked association with *Galearis spectabilis* and maple and tulip poplar trees. They are typically located within close vicinity to the orchid colonies.

Spring flowers that bloom concurrently with the orchids are bear corn (*Conopholis americana*), Jack-in-the-pulpit (*Arisaema triphyllum*), spring beauty (*Claytonia virginica*), and some *Trillium* species.

Explore creeks, ravines, and lowlands that surround the Harpeth River. Although the orchids are extremely small and low growing, the bright colors show up against the dark leaf litter.

Galearis spectabilis blooms in the month of April in Nashville.

Roots

Galearis spectabilis produces underground rhizomes that give rise to small clusters of fleshy roots. The rhizome has a terminal end that will gradually decay as a new bud develops on the opposite side. Rhizomes and roots contain nutrients and carbohydrates for the season's growth cycle. During the months of September and October, new shoots occur on the surface of the soil during seed set. They will be protected from cold temperatures by a thick covering of humus and leaf litter.

Foliage

In March, rising temperatures and sunlight initiate growth. Shoots develop into one or two basal leaves (fig. 6.2). Leaves are broadly elliptic to ovate in shape, 8–20 cm long and 2–10 cm wide. Mature leaves acquire

FIGURE 6.2. A tiny shoot appears in late October as leaves start to disintegrate. The shoot will be protected by leaf litter during the cold winter months.

FIGURE 6.3. A snail travels down the succulent leaf anticipating a great meal. This member of the Mollusca phylum has a radula, a rasping tongue that shreds plant material.

clear margins, a conspicuous midrib, and a wide, blunt apex. Epidermal cells create tiny optical prisms that sparkle in the sunlight, giving the orchid a succulent appearance.

During the first to second week in April, a terminal inflorescence develops between the center of the two leaves. Prominent bracts, elliptic to lanceolate in shape, surround the developing floral buds. Bracts are 15–75 mm in length and 5–15 mm wide. Since their length exceeds the length of the buds, they are frequently mistaken for leaves. On average, three to five tightly closed buds appear. The buds, white to pale pink, intensify in color as they develop. At maturity, the plant body ranges 5–20 cm in height.

The succulent leaves are a favorite food for many insects. Snails, grasshoppers, and planthoppers can eradicate small colonies (see fig. 6.3).

FIGURE 6.4. Two petals and three sepals converge to form the helmet-like hood. Under the hood are the reproductive parts of the orchid.

The Flowering Cycle

The long, slender floral buds begin to open in April. As the buds mature, they develop a helmet-like hood at the apex. This unique arrangement is formed by two lateral petals and three sepals curved downward and converged over the column. The petals are 10-15 mm long and 2-3 mm wide near the base. The sepals are 10-20 mm long and 5-6 mm wide. These range in color from a pale pink to a purplish-pink.

Hidden beneath the helmet is the column. It contains the stigma, style, and stamen, which are the reproductive parts of the flower (see fig. 6.4).

A third petal is modified to form a white, broadly ovate to lance-ovate labellum. It ranges in size from 10-19 mm long and 6.5-13 mm wide.

FIGURE 6.5. The scalloped-edge labellum is as showy as the pink helmet located at the apex of the orchid. Notice that the opening to the nectar spur can be clearly seen.

Lower margins of the labellum vary in appearance; they can be smooth, irregularly scalloped, or fully crenate on the edges.

A distinct white spur, 10–20 mm long, is located at the base of the labellum (fig. 6.5). The meniscus of nectar can be seen inside the translucent spur. The opening to the nectar spur is sized to allow the proboscis of the pollinator. This is a favorite location for spiders. They conceal webs in the nectar opening and lie in wait to prey on various insects.

Despite its small size, *Galearis spectabilis* is exceedingly fragrant. The scent is a fresh citrus-like aroma that is intense in bright sunlight. Pollinators are attracted by the pheromone qualities of the fragrance.

Flowers stay viable on the orchid throughout the month of April, if there is no sudden frost. If successfully pollinated, the ovaries start to develop. If pollination does not occur, the flowers will desiccate on the plant.

Pollination Mechanisms

Galearis spectabilis is morphologically suitable for insect pollination. The labellum with its showy, cuneate margins provides a perfect landing platform for potential pollinators. *Bombus* queens, as well as other bees and flies, are attracted to the appearance of the labellum and the orchid's intense fragrance.

Pollination activity is at its peak on bright, sunlit days. The majority of foraging takes place from mid-morning to mid-afternoon.

The mechanisms for pollination originate under the helmet. Contained within the column are the rostellum, the stigmatic surface, and the enveloping sheaths that hold pollinia. The rostellum is a projection with two rounded prominences that separates the stigmatic surface from the anthers. The stigmatic surface has a sticky gel-like coating that is the location where the pollinia will be deposited in order to initiate fertilization.

When the *Bombus* queen extends her tongue into the nectar spur, the enveloping sheaths above the rostellum pull downward, releasing the pollinia. Subsequently, the queen becomes equipped with pollinia on her head or eye. As she continues to forage for nectar rewards, she deposits the pollinia into an adjacent column. If conditions are suitable, pollination will take place.

FIGURE 6.6. The *Augochlora pura* is actively searching for the nectar rewards.

The orchid maintains a mechanism for the potential of self-pollination. When an insect removes pollinia, it may re-enter the same flower again, thus making it possible for fertilization to occur.[2]

The author observed frequent foraging of members in the Halictidae family, the green metallic bees of *Augochlora pura* (see fig. 6.6). After the bees withdrew from under the helmet of the flower, pollinia was not observed on the eye, head, or body. Whether *Augochlora pura* bees are active pollinators for this orchid was not determined by the author.

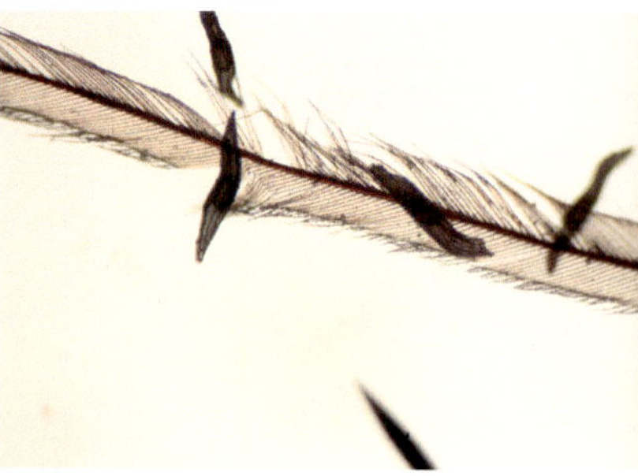

FIGURE 6.7. *Galearis spectabilis* developing seed pods after pollination. The pods contain dust-like seeds that travel by wind currents.

FIGURE 6.8. Seeds from the *Galearis* orchids were discovered on feather samples. Could this be a possible mode of dispersal?

Seed Set, Seed Dispersal, and Wintering

After successful pollination, the inferior ovary starts to enlarge. Over a period of weeks, the ovary will produce a fruiting body (fig. 6.7). At maturation, it is an elliptical to oblong vertical capsule measuring 12.5–25 mm. Each capsule has three or more raised ridges and if cut horizontally, appears in the shape of a triangle. The capsule is green when immature and turns brown when the seeds are fully developed. The capsule splits along the depressions of the raised ridges to disperse numerous seeds.

Typically, the majority of seeds are scattered in the proximate area of the parent plants and flourish in the same vicinity. Air currents, rain, and melting snow can carry the seeds to other locations. Another possibility for dispersal is through animals and birds.[3]

IS AVIAN SEED DISPERSAL POSSIBLE?

In late autumn of 2022, while photographing seed pods of *Galearis spectabilis*, the author noticed small piles of scattered bird feathers 88.5 cm away. Samples of the feathers were collected and examined under a microscope. Several samples held seeds within the matrix of individual feather blades (see fig. 6.8). The author postulated that birds may unintentionally disperse viable seeds through their feathers.

The results of a study in Japan presented evidence of avian dispersal. Birds ate the fruits of the orchid *Cyrtosia septentrionalis*. By examination of the fecal pellets, seeds were shown to be intact and viable.

Thus variations of seed dispersal strategies may be an overlooked mechanism that could exemplify the orchid's degree of advanced evolutionary progress.[4]

Observations on Special Characteristics

LEAVES

The surface of the early spring blooming *Galearis spectabilis* plant tissues have a glistening, granular texture (see fig. 6.9). Bundles of cells catch light and disperse it into tiny prisms. The appearance of the outer epidermis texture is similar to *Liparis liliifolia*, an orchid that blooms approximately a month later. On microscopic view, the epidermal cells are distinctly spherical.

The green-pigmented chloroplasts are the organelles that contain chlorophyll. The principal pigment is chlorophyll *a*, which is found in all

FIGURE 6.9. The crystalline surface of the bracts.

FIGURE 6.10. The leaves contain spherical cells of green chlorophyll.

FIGURE 6.11. Although the cells look transparent, they contain structural proteins within the cytoplasm.

green plants. Contained within the chloroplasts are tiny, closely stacked layers, the grana. The grana layers, or thylakoids, are sites where the light reactions of photosynthesis occur. The cells in many native orchid species are specialized to efficiently perform the photosynthetic process under low light conditions (see fig. 6.10).

THE CYTOPLASM

Figure 6.11 is a microscopic view showing the marginal edge of a *Galearis spectabilis* leaf. The surface cells are rounded with peaks at each apex. Cytoplasm is the transparent gel-like liquid contained within the cells. Cytoplasm was once assumed to be an inert, free-floating liquid. With the invention of the electron microscope, this archaic speculation has been contradicted.

The cytoplasm was discovered to contain a number of cellular organelles, such as the plant's nucleus. Inside the cytoplasmic liquid, structural proteins are distinctly arranged and encased by a ridged cell membrane. Within this protein complex, sites of chemical reactions occur that regulate cellular growth and replication.[5]

FOLKLORE OF

Galearis spectabilis

MY FATHER, C. C. CATES

My father was born in 1917. During his brief life span, he enjoyed telling stories to family and friends, often about the world of nature. This helped instill an active imagination during my formative years. He encouraged me to create stories and songs with my guitar. In the last hours of his life, he advised me to "keep on singing."

One of the few family vacations we experienced was at a location deep in the hills of Middle Tennessee. We stayed in a little cabin near a lake. Since I was from the flatlands of West Tennessee, it was a new experience. We hiked trails and looked with wonder at every plant and tree. We had brought along our Brownie camera and took lots of pictures of the woods and the wildflowers. *Galearis spectabilis* was in bloom at that time, but we did not know what type of wildflower it was. A few months later we decided to identify the plants and trees in the vacation photographs. We were very happy to learn it was a native orchid.

As we sat around a low campfire in front of our cabin, my father asked me to compose a story about those tiny pink flowers we saw on our hike. I began by speaking in a mysterious tone to set the mood. "Those little flowers have helmets on because they are really fairies in disguise. They are trying to hide themselves from the grown-ups," I giggled. "At the stroke of midnight, they will remove their helmets and put star glitter in their hair. They will dance around an enchanted fairy circle of mushrooms until the rooster crows at dawn. Then, the Queen Fairy will proclaim that the dancing be stopped. She will announce that it is time for the fairies to sprinkle magic gold dust on the ground for little girls to find in the morning. The gold dust brings happiness to all the little girls that discover it."

FIGURE 6.12. *Galearis spectabilis* with a visitor, the orbweaver *Theridion frondeum*, who has captured a hearty lunch.

"Well, that was quite a story," my mother said as she laughed and handed out some marshmallows to toast over the campfire.

Next, it was my father's turn to tell his story. He began quietly, "Now, daughter, this is a true story I am going to tell you. When my mother, your grandma, was a little girl, she always sat by the fireplace at night. She claimed she saw a tiny leg sticking out of a hole in one of the fireplace bricks. She just knew it had to belong to a fairy. She ran and told everyone in the town," he said, laughing. "Back then, no one wanted to hinder a child's creative imagination, so they all acted amazed at her story," he added. As he finished the tale, the campfire embers grew dim. Mom stirred the ashes with water to extinguish it. We were ready to turn in for the night.

When we arrived home after the vacation, I looked to see if we had a small hole in our fireplace. To my delight, there was a tiny hole in the limestone brick. So I made it a habit to look for a minuscule fairy leg every night before I was sent off to bed. Alas, I never saw a fairy leg, but I did see some beautiful looking "gold dust" under my swing set one morning. I never found out where the gold dust came from, but I bet it was left by one of those *Galearis* fairies!

FIGURE 7.1. A colony of *Goodyera pubescens* blooming in the summer.

7

Goodyera pubescens

It's Going to Be a *Goodyera*

IF A CONTEST IS HELD for the orchid with the most dramatic leaves in the forest, *Goodyera pubescens* (Willdenow) will take the prize![1] The leaf, with its velvety texture and net-like pattern, is an astounding find in the woods on an early spring day.

The common name is Downy Rattlesnake Plantain (fig. 7.1). The word *downy* describes the pubescent hairs that are abundant on the orchid. The descriptive word *rattlesnake* applies to the tiny, fine veins that form a prominent reticulate design, similar to the pattern in rattlesnake skin. Although the basal rosette arrangement has been noted to resemble the common plantain, it is not in the *Plantago* genus.

This herbaceous, evergreen perennial occurs in central to eastern Canada, stretches into Maine, across to Minnesota and down into Oklahoma with occurrences as far south as Florida.

It is located in a variety of habitats: acidic well-drained slopes and bluffs, moist conifer forest, deciduous woodlands, and dry forested areas.

The orchid has an extensive history. Carl Ludwig Willdenow (1765-1812) was a German botanist who placed the orchid taxonomically in the genus *Neottia*. At that time, *Neottia* included several different orchid genera.[2] A Scottish paleobotanist, Robert Brown (1773-1858), verified the orchid as a separate genus.[3] In 1813, he published a binomial establishing the current name, *Goodyera*, after the esteemed British botanist John Goodyer.[4]

HELPFUL HINT

The striking foliage of this orchid makes it easy to identify. Look for the patterned foliage throughout the seasons of spring, summer, and autumn. Investigate areas within the deciduous woodlands, especially around the trees of sugar maple (*Acer saccharum*), tulip poplar (*Liriodendron tulipifera*), sweetgum (*Liquidambar styraciflua),* paw paw (*Asimina triloba*), pines (*Pinus* spp.), and oak (*Quercus* spp.). When locating this orchid during the colder months, search for the thick, substantial rhizomes that stay green all year long. The leaves may be absent or reduced in the winter due to decreased light intensity.

Goodyera pubescens blooms in the months of July through September in Nashville.

Rhizomes

Goodyera pubescens has sturdy horizontal rhizomes that creep along the surface of the leaf litter and soil. Rhizomes appear as segmented stems that contain outbranches of ascending rootlets (fig. 7.2). These anchor the plant into the soil and stay evergreen throughout the season. Vegetative propagation is achieved by offshoots that develop into new rosettes, typically after the parent plant goes into senescence. Plants are found in colonies or as solitary plants located within close proximity to each other.

FIGURE 7.3. The wet leaves of *Goodyera pubescens* stand out against the dark brown leaf litter in early spring.

FIGURE 7.2. A millipede, a member of the Arthropoda phylum, crawls among the sturdy rhizomes before a rain.

Foliage

Five to eight new leaves unfurl from the central area of the basal rosette. Leaves are alternate, broadly elliptic to elliptic-lanceolate in shape. They measure 2.5–7.5 cm long and 1.3–3.5 cm wide. The leaves are bright green to bluish-green in color. As the growing season progresses, the older, lower leaves become tinged with yellow on the edges.

Occurring at the base of each leaf blade is a white- to cream-colored midvein. It is bordered by smaller parallel veins, four on a small leaf and up to eight on a large leaf. A network of tiny cross-veins runs between them and converges near the obtuse apex. The combination of veins forms the distinct net-like pattern that is a defining characteristic of this orchid (fig. 7.3).

The Flowering Cycle

During the month of July, a cylindrical pubescent stalk, a peduncle, rises from the center of the rosette (fig. 7.4). At the top is an inflorescence that is tightly packed with floral buds. Together, the peduncle and inflorescence span a height of 10–40 cm tall. Along the length of the peduncle are leaf-like bracts, which are densely pubescent.

Tightly closed floral buds appear similar to a cluster of minuscule white grapes (fig. 7.5). Each bud is subtended by a linear-lanceolate bract. Buds open in an ascending manner. On sizable plants, up to fifty or more flowers may be produced.

Comparable with other terrestrial orchids, *Goodyera pubescens* contains three sepals and three petals. The petals are oblong, 3.5–5.5 mm long and approximately 3 mm wide. A dorsal sepal and two petals converge to form a hood over the column. The hood measures 3.5–5.7 mm long and up to 3.5 mm wide. The dorsal sepal is densely pubescent with an upturned apex.

Stigma, style, and a single anther are fused together to form the column. This is the site where pollination takes place.

A modified petal forms the pouch-shaped labellum, which has an apex resembling a tiny down-turned spout. It measures 2.5–4.5 mm long and up to 3.5 mm wide. Two white lateral sepals, ovate to ovate-lanceolate and concave, are etched in green. They flare out on the sides of the labellum. They measure 3.5–5 mm long and 3.5–4.5 mm wide. These form a triangular shape to the flower on frontal view.

FIGURE 7.4. A developing inflorescence.

FIGURE 7.5. *Goodyera* flowers develop in an ascending mode.

An inferior ovary, 5.5–8 mm long, is located beneath the flower. At the base of each ovary are tiny ascending bracts. Pollinated flowers are followed by ellipsoid to oval-shaped fruiting capsules that appear in autumn. The capsules become brown when mature and contain numerous seeds.

Pollination Mechanisms

Goodyera pubescens blossoms have an intense citrus-type fragrance during bright sunlight and in temperatures above 75° F. The metallic green-blue bee *Augochlora pura* was observed to be prevalent at this time and temperature. They foraged the flowers recurrently from mid-morning until noon (fig. 7.6). Other intermittently foraging bees included members of the *Bombus* genus, small bumblebees. Whether they served as active pollinators was not confirmed by the author.

QUESTIONABLE POLLINATORS

Insects forage flowers for several reasons: to seek nectar rewards, find temporary shelter, pursue prey, and display mating behavior. Could these serve as accidental pollinators? The insects described here were frequently observed by the author to have pollinia on their eyes, head, or body.

FIGURE 7.6. A metallic *Augochlora pura* bee forages mature flowers.

FIGURE 7.7. An ichneumonidae wasp forages the *Goodyera pubescens* flowers.

FIGURE 7.8. Two *Verrucosa arenata* spiders await their prey on a mature *Goodyera* inflorescence.

An ichneumonidae wasp occupied crevices in the blooms to catch prey (fig. 7.7). Pollinia was noted on the bodies and wings. By observation, the pollinia was transferred on the surface of the flowers, but not intentionally introduced into the column or labellum. This wasp is mentioned in a book by Henry Baldwin, *The Orchids of New England*, published in 1884. He wrote, "Ichneumonidae at first surpassed all other visitors in observation and discernment, and were thus able to produce inconspicuous flowers which escaped the notice of other visitors."[5] In occasional circumstances, it may be possible that the ichneumonidae wasp, along with other foraging insects, could inadvertently pollinate *Goodyera pubescens*.

This orchid is one of the principal summer territories of the triangular-shaped *Verrucosa arenata*, an arrowhead orbweaver spider (fig. 7.8). The arrowheads would stake out territory on a specific blossom for several days at a time. They built tiny webs between the flowers. They could be seen plucking their strings of silk in order to locate and lure prey. Pollinia was observed attached to their opisthosoma (abdominal region), but not purposefully introduced into the column or labellum.

IMPORTANT NOTE

The mechanisms of orchid pollination cannot be confined to a single theory. Much like the observation methods utilized by Darwin and Thoreau, one must spend protracted hours studying the animal or plant in question in order to comprehend the complexity of the sphere of ecology it encompasses. Modern technology such as motion-detection cameras cannot replace the art of simple, present observation. One must regard the whole environmental picture. Some wildlife activities are so detailed, they can only be interpreted through close inspection by the human eye.

Seed Set, Seed Dispersal, and Wintering

As late summer stretches into early autumn, the green inflorescence changes to brown. Ovaries, located at the base of each dried blossom, begin to swell. Seed capsules start to develop; each one will contain an incredibly large number of seeds (fig. 7.9).

By late autumn, the seeds have matured. The seeds are shaped like long cylinders in a paper-thin sheath. The seed coat is a desiccated shell made up of thin, rectangular cells. An oval embryo is apparent on microscopic view (see fig. 7.10). The embryo may only contain a few cells and most are devoid of any endosperm.[6]

As the capsule splits, thousands of seeds are circulated. Since the seeds are much like fine dust, the slightest air current allows for dispersal. Most of the seeds will fall within the area of the colony. They become scattered among the rhizomes of the parent plants. They lie dormant throughout the winter.

Dried seed capsules persist long after the first snowfall. Much like the seed capsules of *Tipularia discolor*, the dried seed hulls hold water. The author has noticed various insects drinking water from these "natural canteens" during the winter months.

In the spring, seeds begin the long process of development. Fungi are located in the soil at the roots of the plants and trees. The fungi multiply in number, responding to the increased humidity and temperature. Threads of the fungal hyphae enter the cells of the embryo, facilitating nutrients and carbohydrates for growth. The embryo increases in size to form numerous cells assembled into a structure called a protocorm.

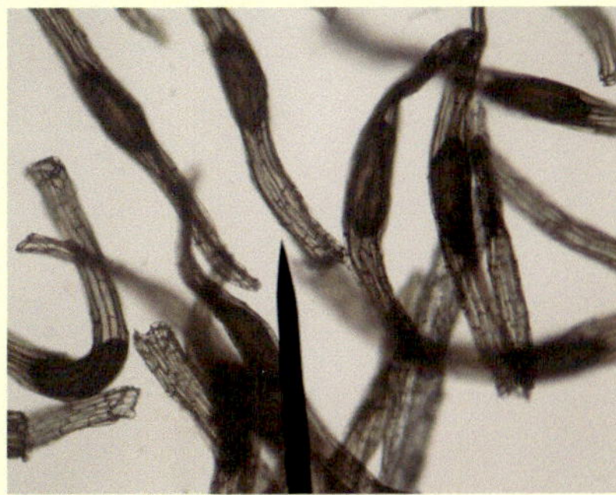

FIGURE 7.9. Newly developed seed pods. Note the spirally twisted ovaries where seeds develop.

FIGURE 7.10. *Goodyera pubescens* seeds have a cylindrical shape. Note the oval embryo in the center of each seed.

Observations on Special Characteristics

THE RELATIONSHIP WITH THE SPINED ORBWEAVER

Micrathena gracilis (Araneidae) has an advantageous relationship with *Goodyera pubescens*. This spider is commonly called the spined orbweaver or the castle-back orbweaver due to its bulbous, spiny-looking abdomen (fig. 7.11). Only the females exhibit this characteristic. Males are much smaller and seldom seen except when mating. Colors can vary from brown to black on head and legs with white strips on the abdomen.[7]

FIGURE 7.11. The spined orbweaver, *Micrathena gracilis*, waits patiently for prey at the apex of the bud.

FIGURE 7.12. A leafhopper in the family Cicadellidae inflicts wounds that expose *Goodyera pubescens* buds to fungal rot.

The spined orbweaver occupies habitats with humidity and moisture, such as creeks, ponds, and the moist deciduous woodlands that surround the city limits of Nashville.

Large wheel-like webs are built in open areas such as hiking trails to take advantage of natural fly-ways. Insects travel these fly-ways and become easy prey for the orbweaver. Attached to their web is a single delicate line of silk. This allows the orbweaver to maneuver from the web to an adjacent orchid bud or a tree in a manner of seconds.

Spiders play an important role as biological pest control agents.[8] They are efficient in capturing harmful leafhoppers, planthoppers, and treehoppers that can cause devastation to the entire plant (fig. 7.12). Feeding on the plant juices, hoppers serve as vectors of plant viruses. They also harm the plant by inflicting wounds that expose the orchid to fungal attacks. Certain bacteria, fungi, and viruses can eradicate whole colonies of *Goodyera pubescens*. More studies are needed to determine the importance of spiders as control agents against opportunistic pests.

FOLKLORE OF

Goodyera pubescens

Early settlers made pastes, poultices, and tonics from the various wild plants collected near their homes. The leaves of this orchid, as with other terrestrial orchids, were used for conditions such as hives, abscess wounds, and snake bites.[9] It must be remembered at this time in American history, medical facilities were not available. Doctors were few in number and many miles away from settlements.

MR. H. CALLDAL

The late Mr. H. Calldal related a tale to me about the use of the "rattlesnake plant" for various ailments.

He began, "Now your great-grandaddy knew about these plants too, and we used them on the farm as they grew there in the woods. The leaves were used when the old folk had kidney problems or when the children kept getting up in the night to go to the bathroom. Back in those days, most folks didn't have an indoor bathroom, they used an outhouse. It could get real cold in the winter out there, if you know what I mean," he chuckled.

"Sometimes the old folks would get the pneumonia and the kids would too. The women folk would make a poultice out of the rattlesnake leaves. I think I recollect how they did it. The leaves were mashed and mixed with boiled vinegar. There were some other ingredients but I don't remember what they were. They secured it, the poultice that is, in a pig or sheep bladder and placed it inside a wool bag. The bag was kept hot and placed in the areas of the lower back of the person with kidney problems. When folks had the pneumonia and croup, the leaves were heated together with mule manure. Your great-grandaddy raised lots of sturdy mules back then. Well anyway, these leaves were put in a canvas bag and placed on top the chest. They had to turn the sick ones every

FIGURE 7.13. This beautiful specimen of *Goodyera pubescens* was located a few meters away from a busy highway.

few hours and kept on heating the poultice or it wouldn't work right. This took an awful amount of labor and time, but it usually made the folks well and on their feet in a week or so. The secret ingredient was claimed to be the mule manure. It generated heat for the poultice to work its healing properties."

He reminisced, "I was lucky 'cause I never got the pneumonia or the argue. I never got sick much and neither did your great-grandaddy. That is 'cause we worked on the farm in all kinds of weather. We never even got a cold," he concluded with a hearty laugh.

Mr. Calldal said that he knew of some "old-timers" in town that still believed in the power of the rattlesnake plant.

FIGURE 8.1. The colorful and intriguing *Liparis liliifolia*.

8

Liparis liliifolia

The Celebration of a Spring Orchid

AN EXCITING WAY TO USHER in the spring season is to locate the gentle, curving leaves of *Liparis liliifolia* (L.) L. C. Rich. ex Lindley.[1] The lush, fleshy leaves, spidery inflorescence, and "helicopter pad" labellum are a surprising spectacle to behold among the browns and grays of the forest (fig. 8.1).

This orchid has various common names: Lily-Leaf Twayblade, Large-Leaf Twayblade, Mauve Sleek-Wort, and Purple Twayblade. The Old English word *tway* means two. This refers to the orchid's two basal leaves. The epithet *liparis* is derived from the Greek word *liparos*, which means fat or greasy, describing the extra shiny appearance of the leaves.

The orchid is prevalent in eastern Canada, from Ontario to Quebec, across to the Great Lakes region, and down into the Ohio and Upper Mississippi River Valley. It is rarely found in Florida and Louisiana.

Liparis liliifolia is located in a variety of habitats including mature deciduous woodlands, dry hickory-oak forests, moist woodland slopes, open pine forests, and thin, acidic rocky soils. Frequently, it is found in disturbed locations, such as areas of early reforestation.[2]

Populations do not tolerate heavy shade and if exposed to these conditions are inclined to significantly decrease in number.[3] Newly reforested areas and reductions of underbrush help to maintain existing orchid communities by allowing light into the forest canopy.

HELPFUL HINT

These orchids can be found in many of the parks and greenways in Metro Davidson County. Locate the trees of sugar maple (*Acer saccharum*), hickories (*Carya* spp.), American beech (*Fagus grandifolia*), eastern red cedar (*Juniperus virginiana*), sweetgum (*Liquidambar styraciflua*), tulip poplar (*Liriodendron tulipifera*), and oaks (*Quercus* spp.). Other areas to explore include open pine forests and cedar thickets. Search under wild grapevines and fallen trees.

During the second week in April, look for the emergence of a vase-like pair of leaves. They will appear bright green against the dark humus and soil. To locate the orchids in winter, look for the sheathing dry scales, the remnants of dried inflorescences, or uprooted leaf bases. Since they are shallow rooted, their locations can shift throughout the season.

Liparis liliifolia blooms in the month of May in Nashville.

Description of the Corm

Liparis liliifolia orchids are produced from two fleshy, dense, ovate corms that have overwintered from the previous growing season (fig. 8.3). Also referred to as pseudobulbs, they range in size from 12-20 mm in diameter.

In this guide, the two corms will be designated as parent and daughter.[4] Each corm is encased in several sheathing, paper-like scales. Frequently, the parent corm has a dried leaf base and stalk still attached

FIGURE 8.2. A tiny white-footed mouse, *Peromyscus leucopus*, enjoys seeds as well as the *Liparis liliifolia* corms in the autumn and winter.

FIGURE 8.3. A new orchid springs into life from the attached corm. Notice the dried inflorescence is still attached to the parent corm.

from the preceding season. The parent corm is connected to the daughter by a finely gauged rhizome. Both parent and daughter corms produce shallow, fibrous root systems. The daughter corm contains stored carbohydrates and nutrients for the current season's growth. Corms, in various stages, may persist on the orchid for several years.

Corms are shallow and easily uprooted by strong rains and snow melts. In this manner, they are transplanted to a separate location away from the parent plants. If conditions are favorable, they will begin to establish new growth.

Wildlife such as the Eastern cottontail (*Sylvilagus floridanus*), the white-footed mouse (*Peromyscus leucopus*; fig. 8.2), and the white-tailed deer (*Odocoileus virginianus*) enjoy browsing the corms. This can lead to the destruction of plant populations. However, in recent years, runoff from the residential use of herbicides and insecticides has destroyed plant communities along with their numerous pollinators.[5]

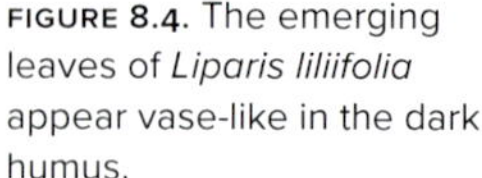

FIGURE 8.4. The emerging leaves of *Liparis liliifolia* appear vase-like in the dark humus.

FIGURE 8.5. Glossy, succulent-looking leaves are characteristic of *Liparis liliifolia*.

FIGURE 8.6. Floral buds look like a tiny green human hand reaching upward into the sky.

Foliage

Liparis liliifolia leaves emerge from the dark humus and soil during the month of April. The daughter corm gives rise to two elliptic to ovate, light to deep green, succulent-looking leaves. These are evenly keeled abaxially at the base of the corm. A sterile plant will produce a single leaf.

Young developing leaves are erect, with a vase-like appearance (see fig. 8.4). The outer margins of the leaves are smooth with an obtuse apex. There is a conspicuous mid-rib with pale, parallel veins. On maturity, there is great variation in the size of the leaves depending on the nutrient content and girth of the supporting corm. The leaves flex downward, approximately 3.5–18 cm long and 3–8 cm in width (fig. 8.5). Mature plants vary in height, from 8–25 cm.

Leaves remain green and succulent on the orchid during bloom time and throughout seed set. In darker forest canopies, leaves may remain on the plant until late autumn, while leaves in open areas will decay much earlier. After senescence, the leaves disintegrate into a straw-colored matrix.

In May, floral buds start to appear in the center of the conduplicate leaves. Developing buds are green and erect with enclosed margins. Optically, the buds can take on some surprising human characteristics. The author spotted buds with physical expressions much like that of human hands. Some were positioned as in prayer, while others were reaching upward toward the sky (see fig. 8.6).

The Flowering Cycle

The flower morphology is complex with its insect-like appearance. Buds start to open in an ascending mode, from bottom to top. They are supported on bright purple-tinged pedicels measuring 4.5–7 mm long. Each pedicel has a tiny bract at the base. As mid-May approaches, a six-sided, segmented, winged inflorescence arises from the center of the leaves. The inflorescence is a loose raceme that will yield prominent mauve flowers. Up to thirty pendant flowers may be produced on robust plants (fig. 8.7).

An oblong-lanceolate dorsal sepal is located directly above the column. It appears spear-like and measures 8–11 mm long and 1–2 mm wide. The winged column resembles a tiny insect head or a miniature bird beak that arches above the labellum measuring 2.5–4 mm long and 1–1.5 mm wide. It contains the reproductive parts of fused pistils and stamens. Two tiny tubercles are located on the inside surface near the column's base.

FIGURE 8.7. A spectacular example of a mature *Liparis liliifolia*.

The two lateral petals are curved and filiform-shaped. They appear as tiny strands of purple thread and account for the spidery appearance of the orchid. They range in size from 8–12 mm long and 0.2–0.3 mm wide.

The mauve labellum is a conspicuous, translucent disk measuring 8–12 mm long and 6.5–10 mm wide. On closer inspection, it has a delicate network of reddish-purple veins running downward toward the apex. The transparency of the labellum allows two lanceolate, light green sepals to be seen underneath. Each sepal measures 8–11 mm long and 1–2 mm wide. The tips of the sepals can be seen as if they are "peeking out" on either side of the base of the labellum.

In the center of the labellum is a pale, glistening, slightly constricted strip. It appears as a dividing line down the middle of the labellum and attracts insects toward the column for potential pollination. Ants are commonly seen foraging on the strip. The frequency of their appearance suggests that there may be nectar rewards.[6] Whether the strip exudes nectar or a pheromone remains to be seen.

Flowers can remain up to three weeks on the inflorescence.

Pollination Mechanisms

Liparis liliifolia is self-incompatible, as it requires cross-pollination to produce viable seeds.[7] Since the flowers have the coloration and veins that resemble carrion, it has been suggested that various small flies are active floral pollinators.[8]

The flowers have a definite carrion odor, much like spoiled meat. The scent is prominent during sunrise and nautical twilight. Fruit flies (Tephritidae), mosquitoes (Culicidae), and midges (Chironomidae) were observed foraging the flowers during these hours (see fig. 8.8). Although these flies were very small, they would brush against the column where occasionally the pollinia would attach to the head and wings. The pollinia in relation to the insect body was prominent. March flies (*Penthetria heteroptera*) were frequently seen in the years 2020 through 2022. The flies foraged the flowers and engaged the labellum as a suitable location

FIGURE 8.8. *Liparis* with two foraging flies on the shiny central strip of the labellum.

FIGURE 8.9. *Penthetria heteroptera* during mating. Notice the pollinia attached to the wing of the fly on the left.

for mating purposes. The author noticed pollinia attached to the body and wings, yet it was not observed to be transferred into the columns. However, the possibility of the flies serving as passive pollinators cannot be ruled out (see fig. 8.9).

Butterflies and moths of the Lepidoptera order are attracted to these orchids. The brilliant swallowtail, *Papilio glaucus*, is often seen landing on the labellum. Whether it may be attracted to the color of the flowers or encouraged by obtaining nectar rewards is not known. (See fig. 8.13 in the Folklore Section).

Seed Set, Seed Dispersal, and Wintering

By late summer, if successful pollination occurs, the orchid produces fruiting capsules. Bright green and winged, the capsules are supported on pedicels that are as long or slightly longer in measurement than the capsules themselves (fig. 8.10).[9] Pedicels measure up to 18 mm long in comparison to the elliptic fruiting capsule measuring 15 mm long. The capsules hold a vast number of immature seeds. When the seeds are fully developed, the fruiting body starts to change into a desiccated,

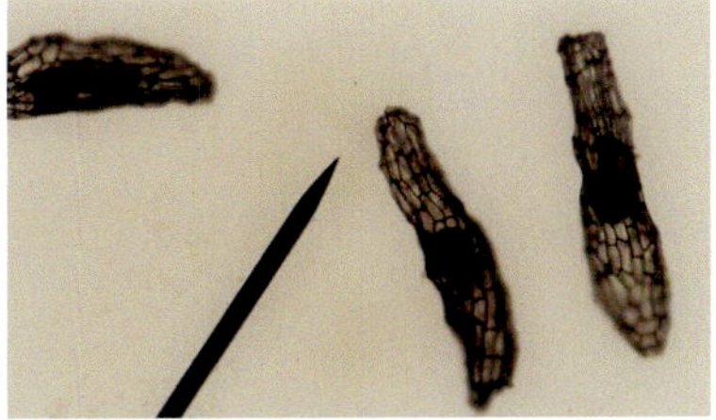

FIGURE 8.10. Winged-seed capsules are located on long pedicels. These hold a vast number of seeds. Notice that the leaves are still vibrant and green.

FIGURE 8.11. Microscopic view of *Liparis liliifolia* seeds. The oval embryo is in a central location surrounded by a net-like seed coat.

brown capsule. Both the dried capsules and the inflorescence remain throughout the winter season.

Atmospheric moisture will signal the seed capsules to split open and disperse the seeds. Most seeds fall in proximity to the parent plants. Other seeds are scattered by wind currents, rain showers, and snow melts. In this manner, seeds can be transported for long distances. These methods of dispersal can help in establishing new orchids in another suitable habitat.

After the seeds have been dispersed, they settle in the humus and lie dormant during the winter. Germination occurs with rising temperatures and the symbiotic relationship between fungi and orchid. As with most terrestrial orchids, the fungi infiltrate the seeds that aid the embryo with nutrition for growth.

On microscopic view, the dust-like seeds are more orbicular in shape in comparison to the seeds of other terrestrial orchids that grow in the same vicinity such as *Galearis spectabilis*, *Goodyera pubescens*, and *Tipularia discolor*. The net-like seed coat has large air spaces with a dense oval embryo located in the center (see fig. 8.11).

Observations on Special Characteristics

A DOUBLE FLOWER

Occasionally, nature creates a flower that is unusual and noteworthy. In 2021, a curious *Liparis liliifolia* flower (fig. 8.12) appeared on a typical inflorescence with twenty-two adjacent standard-sized flowers. It developed two columns with accompanying pollinia, two completely formed labellums, and two dorsal sepals on one pedicel. Each labellum was attached individually at the base where the sepals arose. However, three sepals were observed under the labellum, instead of four as in a perfect duplicate.

This flower was preferred by foraging insects. Insects were observed to challenge each other for their right of territory over this duplicate flower. The sequence of events began when an insect would land on one side of the double labellum. Next, it would examine the flower for nectar rewards. If an approaching challenger tried to land on any part of this flower, the principal insect would charge toward this intruder with their antennae raised. If this tactic did not work, it would aggressively fly into the challenger and push it off the flower. The challenger would retreat and move on to conquer another flower out of range of the aggressor.

THE DOUBLE "I" IN THE NAME EXPLAINED

In 1753, Linnaeus first described the orchid as *Ophrys lilifolia*, with a single "i" in the spelling. In 1800, biologist Olof Swartz used the double "i" in the species and placed the orchid in the genus *Malaxis liliifolia*. This started a series of debates that ensued for many years. Later, Lindley categorized the species into the *Liparis* genus, naming it *Liparis liliifolia* (note the double "i" here). All the synonyms now contained the double "i" due to his citing.[10]

Years later, taxonomists began a new debate over the naming of the species. For several more decades, the controversy continued until the

FIGURE 8.12. The double flower with the double "i."

botanist Bernard Boivin determined the outcome in his article in the 1967 issue of *Rhodora*. He corrected the spelling of several species, including *Liparis liliifolia*. He explained in his editorial why it was important to adhere to Article 73, Note 2 of the *International Code of Botanical Nomenclature*, which describes the use of uniformity when applying scientific botanical names.[11]

FOLKLORE OF

Liparis liliifolia

MRS. DEWALT

Years ago, I was shown my first *Liparis liliifolia* orchid by an aged friend, Mrs. Dewalt. She called it a "space plant." She lived on a farm that bordered several acres of deciduous woods in Tennessee. The farm had been in her family for 125 years.

She narrated interesting tales about herbal cures that were used at a time when there were no antibiotics available. She recalled how one man helped to save many children in her community. She began her story: "There was a man who knew the ways of the Cherokee tribe, in curing with herbs and potions. He would travel by horse and buggy down the creek beds to my town and to the many small rural communities around these parts. He remedied many children and delivered jars of homemade ointments and poultices."

She accurately explained his methods. "To cure chest congestions and pneumonia, he mixed some dried space plant leaves [*Liparis liliifolia*], and added young turkey leaves [*Verbascum* spp.], and cocklebur root [*Xanthium strumarium*]. He mashed them all together. The space plant leaves added a sticky quality to the poultice. He used cow manure and finely sliced, fermented onions. Boy, did it have a smell!" she laughed. "The ingredients were mixed, put into bell jars, and set out on the roof in the bright sun. Then it would ferment and then the mixture was put into a bag. We didn't have plastic back then so we used a bladder of a pig or sheep. The healer would wrap the bag onto the chest, back, and neck area of the sick person. Back then, no one knew why he used those ingredients. They just did what he said. It was hot, stinky, and slimy, and we hated it, but we knew it was the cure." She laughed again. "It was hard work 'cause the patient had to be turned several times a day and during the night. The 'cure' took four to six days. It even worked for me!" she said in amusement. Then she added, "To keep the children from picking these plants, they were told that they were planted by goblins or space men."

On a late spring evening you may see an unearthly orchid that looks like it is from another planet. Who knows? It might even be from Venus!

FIGURE 8.13. A swallowtail, *Papilio glaucus*, visits *Liparis liliifolia*.

FIGURE 9.1. A robust *Spiranthes cernua* in early October.

9

The *Spiranthes cernua* Complex

A Taxonomical Puzzle

JUST AS THE FILM DIRECTOR ALFRED HITCHCOCK achieved the title the Master of Suspense, the *Spiranthes cernua* (L.) Richard complex should be granted the Master of Perplexity.[1] Esteemed scientists throughout the ages have disputed the details of the orchid's evolution, geographic origin, morphological variability, and suspected hybridization within the species (fig. 9.1). Members exhibit overlapping variations, creating a challenge for taxonomists to identify and name. Thus, conservation efforts have been difficult to define.

An in-depth investigation was conducted by Pace and Cameron in 2017. They skillfully charted nucleotide alignments and studied morphological variations through field and herbarium samples. The results of this research clarified species boundaries. Presently, *Spiranthes cernua* is split into five distinct species.[2] It is beyond the scope of this guide to define species delineation in detail, so we will refer to this autumnal flowering orchid as *Spiranthes cernua*.

The common names are Ladies' Tresses, Fragrant Ladies' Tresses, and Nodding Ladies' Tresses. Much like the appearance of braided hair, the axis of the inflorescence is twisted into a spiral-like arrangement. This arrangement is a consequence of uneven cell growth.[3]

Spiranthes cernua is endemic to central and eastern Canada, across the Midwest to Kansas and Nebraska, running through the southern Appalachian Mountains, and into the southern states, including Louisiana and Florida.

In the Central Basin, it is found in moist to dry moderately acidic soils. Wide-ranging habitats include bogs, ditches, meadows, grassy fields, older cemeteries, roadsides, and suburban lawn borders.

Of special note is the Southern Slender Ladies' Tresses (*Spiranthes lacera* var. *gracilis*). In contrast to *Spiranthes cernua*, this orchid blooms in late summer, is glabrous, and has the appearance of a spiral staircase. It is mentioned here for comparison, as the author found it prevalent in outlying counties. (See fig. 9.12 in the Folklore Section.)

HELPFUL HINT

Spiranthes cernua is frequently found in suburban lawns. Look carefully as the leaves can be mistaken for grasses. Early mowing in April helps to promote aeration and aids in eliminating competitive vegetation. As the summer progresses, restricted mowing practices encourage growth.

Spiranthes cernua blooms in the month of October in Nashville.

Roots

Roots are variable in shape and size, from slender and thread-like to fleshy and tuberous. They appear brown to yellow in color (fig. 9.2). The fine roots spread horizontally on the surface of the soil, while the tuberous roots descend much deeper. Tuberous roots range from 1-1.2 cm in

FIGURE 9.2. The new shoot of *Spiranthes cernua* emerges. The shoots overwinter until light and temperature initiate growth.

FIGURE 9.3. A synsacrum of a bird of prey stands guard over the new grass-like leaves.

diameter. Roots start to initiate small shoots soon after seed dispersal in autumn. Tiny spike-like shoots appear next to the dried inflorescence of the previous growth season.

Foliage

The spiked shoot emerges to support two to five upright grass-like basal leaves (fig. 9.3). The leaves are elliptic, linear-lanceolate to lanceolate or oblanceolate in shape, with a conspicuous midrib. They are much longer than their width, measuring 5–23 cm long and 5–7.5 mm wide. Clasping bracts, pointed at the apex, occasionally occur below the inflorescence. Plants occur in sizable colonies or as solitary plants located a few meters apart.

During anthesis, some leaves are usually present, although tattered leaves located on the lower portion of the peduncle may be the only ones remaining.

IMPORTANT NOTE

There are numerous keys available to aid in the identification of this species, but a number of these have contradictions as to whether leaves are absent or present at anthesis. In identifying any species, it is important to regard more than one factor. Look for three defining features to make an accurate decision. Some suggestions are to observe flower morphology, habitat preferences, and the occurrence of polyembryonic seeds. Fragrance is subjective, and may not generate valid results for the purpose of identification.

The Flowering Cycle

A vertical inflorescence, 10–50 cm tall, makes its appearance in the latter part of September. It holds twenty to forty loosely spiraled, white, lanceolate blooms. An elongated ovate-lanceolate floral bract subtends each flower. The petals, sepals, and labellum form a pubescent, tubular-shaped flower that appears as if it is nodding downward, hence the common name of Nodding Ladies' Tresses (see fig. 9.4).

At the apex of the flower is a convex dorsal sepal that is recurved upward toward the tip. It measures 5.5–11.5 mm long and up to 2.5 mm wide at the base. Two dorsal petals, also recurved at the tips, converge with the dorsal sepal to form a curved hood. Each petal measures 5.5–11.5 mm in length and 2.5 mm wide.

The flower has two lateral sepals that are linear-lanceolate in shape, each measuring 5.5–11.5 mm long. The length of the sepals sweeps upward above the hood. This is a defining characteristic for *Spiranthes cernua* identification.

The oblong to ovate labellum is recurved downward and slightly constricted in the middle. Crystalline white with a pale-yellow center, it measures 7–12 mm long and 3–5.5 mm wide. The distal portion of the labellum

FIGURE 9.4. This stately inflorescence measured over 30 cm, while the total height of the plant measured 65 cm.

FIGURE 9.5. *Spiranthes cernua* in flower. Notice the *Ceratina* bee flying to a newly opened blossom.

has crenulate, flared margins (fig. 9.5). Overall, *Spiranthes cernua* has larger, more open flowers than others in the split-species complex.

The fresh fragrance of this terrestrial is somewhat difficult to define. Sheviak described it as a coumarin odor.[4] The author, however, noted it was much like a combination of salty sea air and faint suntan lotion, extremely clean and potent. The fragrance was most intense during the mid-morning to afternoon hours.

After pollination, the pubescent ovary, 3–10 mm, swells with developing seeds. A mature seed capsule is produced. The capsule holds innumerable amounts of viable seed.

Pollination Mechanisms

Sunny October days usher in peak foraging activity. Simultaneously, when the sun reaches maximum brightness, the fragrance of the orchid is at its height of intensity. On overcast days, the scent was inconspicuous, thus rendering pollination intermittent. The bees, *Ceratina* and *Bombus* species, are active pollinators. They carry pollinia on their tongues, and occasionally it becomes attached to the eyes, head, or thorax.

The small *Ceratina* species of bee is an important pollinator for commercial crops and native plants.[5] It belongs to the Apidae family of bees, with over three hundred fifty species including the cuckoo, digger, and honey bees. Some bees have a distinct cream- to white-colored heart-shaped marking on their heads. They are commonly referred to as carpenter bees, as they make their nests in the pith of woody stems or rotten logs. They are similar in size, shape, and metallic coloring to bees in the Halictidae family. The *Ceratina* bees observed in active pollination measured 5-6.5 mm in length and had a metallic bluish-black sheen.

Exacting in their choice of floral selection, the bees begin by approaching mature flowers first on the lower portion of the inflorescence. They quickly work their way up toward the newly opened flowers at the apex.

FIGURE 9.6. Close-up of the *Ceratina* bee. Notice the extracted pollen.

FIGURE 9.7. A small bumblebee is observed to be an active pollinator.

The *Ceratina* bee exercises dexterity when gathering pollinia as it enters the flower upside down (see fig. 9.6). As the bee's body completely disappears into the center, the bee's proboscis reaches into the column and collects pollinia. This reversed position apparently helps to ensure that pollinia will come in contact with the bee. When the bee enters another flower, the pollinia is transferred.

The *Bombus* genera of bees are skilled at cross-pollination. Their long tongues can reach the nectar at the base of the floral tube. The pollinia attaches readily to the flat, ridged tongue. In contrast to the *Ceratina* bees, the author noted that the bumblebees had no specific method in choosing flowers to forage. Without any consideration to the degree of maturation of the flower, these bumblebees foraged randomly (fig. 9.7).

The author noticed when the temperatures were greater than 73° F, a battle of wits would occur between the two bees, *Bombus* and *Ceratina*. As competitors for nectar rewards, a sturdy *Bombus* bee would invade the flower that a *Ceratina* bee was actively foraging. The intrepid *Ceratina* would not submit to the challenger. A chase would ensue for a claim over the preferred territory. The victor would reap the rewards of nectar from the choice flower. Many times, the little *Ceratina* would claim the prize as the *Bombus* flew off in a huff!

Seed Set, Seed Dispersal, and Wintering

In late autumn, the seeds have matured in their capsules. The seed coats contain air pockets that contribute to the effectiveness of air and water dispersal.[6] Seeds are dispersed at intervals due to air currents, rain, and the movement of animals over the territory. The dried brown capsules

FIGURE 9.8. This inflorescence contains seeds that are starting to develop within the ovaries. By early winter, they will be fully matured and ready for dispersal.

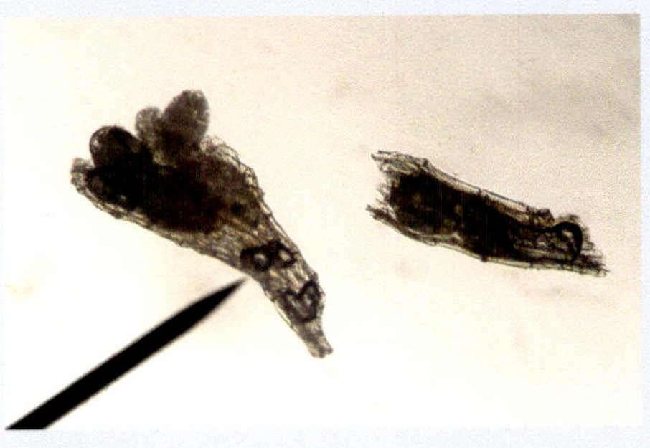

FIGURE 9.9. Mature seeds showing embryonic variation. Note the seed on the left has multiple embryos protruding out of the seed coat. On the right, the polyembryonic seed contains two embryos.

retain their twisted appearance and remain on the inflorescence long after the seeds are distributed (fig. 9.8).

As with most terrestrial orchids, seeds are germinated in close proximity to the parent colony. Germination must be initiated by favorable conditions of climate and light, in addition to the mycorrhizal fungal association with the orchid.

The reproduction of sustainable seed is accomplished by both asexual and sexual behaviors.[7] An interesting parallel is that the *Ceratina* bee and the *Spiranthes cernua* orchid have equivalent adaptations for the mechanism of asexual reproduction. *Spiranthes cernua* has the capacity to carry out agamospermy, which is the formation of seeds without fertilization.[8] In comparison, species of *Ceratina* bees are capable of being parthenogenic; they have the ability to reproduce without males.[9]

A defining characteristic of *Spiranthes cernua* is a mixture of embryonic variation occurring within a single seed capsule (fig. 9.9). Microscopic observation allowed the author to describe the following five variations in the agamospermous seeds: (1) a single, oval embryo of the monoembryonic seed; (2) two oval embryos of the polyembryonic seed; (3) seeds with protruding embryos developing outside the seed coat; (4) seeds with no apparent embryos within an empty seed coat; and (5) free-floating embryos.

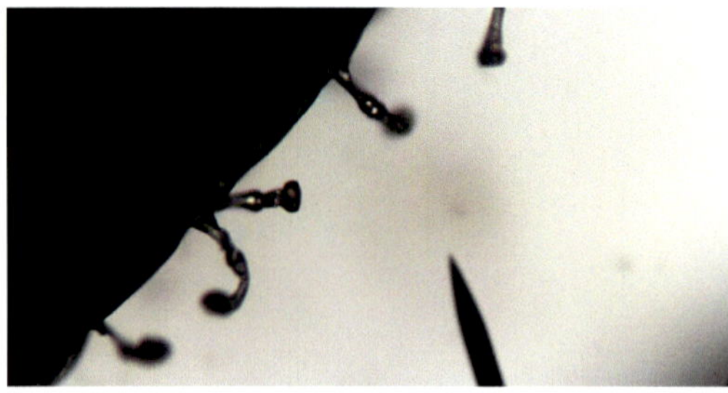

FIGURE 9.10. This microscopic view reveals glandular hairs located on the outside surface of the dorsal sepal.

Observations on Special Characteristics

PUBESCENT HAIRS, OR TRICHOMES

Dense pubescent hairs, sometimes referred as trichomes, cover all parts of the *Spiranthes cernua* flower and the upper portion of the inflorescence. The segmented glands are arranged on stalks with blunt tips (fig. 9.10). Glandular hairs originate in embryotic cells and develop on the outer epidermal tissues at maturity.[10]

Pubescent hairs create a dense covering that helps protect against the sun's excessive radiation. This reduces transpiration by preventing water loss during arid, hot conditions. Hairs function by increasing the surface area to enhance the absorption of water from the humidity in the surrounding air. They discourage pests from attacking the orchid's outer tissues by providing a sticky barrier. Hence, hairs are another example of the orchid's advanced evolution.

THE RELATIONSHIP WITH THE AMBUSH BUG

Within the family Reduviidae is a unique insect commonly named the ambush bug (fig. 9.11). This insect has the ability to capture prey much larger than itself. The thick forelegs with their claw-like tips resemble tiny mantis legs. As a distinguishing trait of this group, they have suitably named raptorial forelegs.[11] In comparison to the rest of the body,

FIGURE 9.11. Perched atop a *Spiranthes cernua* flower, this ambush bug waits patiently for his prey.

the abdomen is large. Responsive eyes on either side of the head swiftly notice any small movement on the orchid flower. The two antennae are clubbed at the ends.

Adorned in a variety of colors, they have the advantage of camouflage. With the ability to change colors, they can blend in with the flowers they occupy or with the background surrounding their host plant.

The insect has a robotic-like locomotion that resembles a child's Transformers toy as it punctuates each movement with a pause and an upward stationary jump. They perform this stealthy movement in all directional turns and when capturing prey.

The individual in the photograph (fig. 9.11) was perched on the largest inflorescence within a colony of six plants. It occupied this position for five consecutive days. This location provided the greatest amount of foraging activity.

This specimen was 6.5 mm long. It was identified daily by a slight scar on the left foreleg. It enjoyed being photographed at close range as it smiled for the camera!

FOLKLORE OF

the *Spiranthes* Genus

MS. J. VANARDY

The late Ms. J. Vanardy related a story about using *Spiranthes* as a hair wash.

She began her story on a bright sunny afternoon. "There were tribes that lived on this land long ago. They had lots of know-how of the plants around them. They were grateful to the Great Spirit. He gave them the plants. If they didn't do what he told them, he would get mad. But he protected them and gave them all the things they needed like food and help when they was sick. They had a medicine man that treated 'em when they were sick. They had their grandmothers to make a magic medicine bag, too. My daddy told me the medicine bag was really powerful and no one was suppos' to touch it but the one it was for. They put lots of plants and other things like rocks and hair from animals in those bags.

"Those 'twisted hair plants' was like those scouring pads you buy down at the store. It was rough and it could clean your insides out, too, if you drank it. I never drank any but ol' Jess did and it made him sick for seven days and seven nights. Folks 'round here thought he was gonna die. But he pulled out of it.

"When it come fall, folks would go lookin' for those twisted hair plants and pull them up just before they were all bloomed out. They were going to use them to make their hair grow long and purty. They would mash them up with rocks and mix them up with water and vinegar and put them into clay jugs. Night came on and they would set 'em out in those jugs in the moonlight. Then they said prayers. Prayers make the potions work better. Sometimes they would sing to pass the time, and my mama said that her great-great-grandmamma could hear them singing into the night while they made the hair tonic. Well, anyway, they set it out for about four to five days and it soured. It smelt bad. The girls knew it was

FIGURE 9.12. The spiral *Spiranthes lacera* var. *gracilis* in August.

going to make them pretty more than they were. They put the mix on the girl's hair and rubbed it in. It had a sort of foam up. The girls slept with it on their heads and in the morning, they washed it out in the cold creek water. Next thing you knew was their hair would grow long down their backs and it was shinin' like the sun. Today, we have to use the stuff from the store, but it don't work as good," she concluded.

After hearing this story, I wished I could have washed my hair in the mixture too!

FIGURE 10.1. *Tipularia discolor* in civil twilight.

10

Tipularia discolor

The Elusive Pixie of the Orchid World

IMAGINE WALKING THROUGH THE WOODS on a late summer afternoon. The light begins to fade as you notice something peculiar out of the corner of your eye. Could it be a strange plant with wings, or is it a circle of tiny flies suspended in air? To your surprise, it is the elusive pixie of the orchid world, *Tipularia discolor* (Pursh) Nuttall (fig. 10.1).[1]

On further inspection, you observe that the "wings and tiny flies" are asymmetrical, long-spurred flowers located on the axis of an inflorescence. The appearance of hovering flies is the source of the common name, Crane-Fly Orchid. The name originates from the family of crane flies, the Tipulidae.

With its solitary green leaf appearing in early autumn and subsequent flowering in the summer, this native orchid plays a game of hide and seek throughout the year. It produces flowers in the late summer when few wildflower species are in bloom.

Tipularia discolor is prevalent from upper Michigan to Massachusetts, sweeping down into the Midwest across to Missouri, west to Oklahoma and Texas, and spanning into the southern states, including Florida. It occurs throughout the state of Tennessee in coniferous and deciduous woodlands. It is the only member of this genus found in North America, while others are located in Asia.[2]

Communities of *Tipularia discolor* exhibit variable patterns ranging from condensed populations to solitary plants located several meters apart. They occur in various habitats within the city limits of Nashville: in layers of thick leaf litter, on moist slopes, near creeks, and inside rotten logs.

HELPFUL HINT

Explore areas around the deciduous trees of sugar maple (*Acer saccharum*), hickories (*Carya* spp.), sweetgum (*Liquidambar styraciflua*), tulip poplar (*Liriodendron tulipifera*), and the oaks (*Quercus* spp.). Search in cedar-dominated woods where the eastern red cedar (*Juniperus virginiana*), common hackberry (*Celtis occidentalis*), and winged elm (*Ulmus alata*) thrive.[3] Take special note of tulip poplar and sweetgum, as the orchid is predominant in these areas, presumably due to the mycorrhizal association provided by these specific trees.

It is easy to recognize this orchid during the winter months, as the solitary green leaf stands out against the browns and grays of the forest. Recording the location during this time of the year will provide effortless identification during the summer season.

Tipularia discolor blooms in July through late August in Nashville.

Description of the Corm

The corm is a basal, bulb-like storage organ that contains nutrients for the developing orchid. On average, corms measure 5–25 mm in diameter, although robust plants can produce corms that are several millimeters larger. Corms are produced at the base of the leaf.

FIGURE 10.2. This happy little snowman is formed from *Tipularia* corms on the surface of the soil.

FIGURE 10.3. White-tailed deer (*Odocoileus virginianus*), a young buck, forages along the deciduous border in the suburbs of Nashville.

Large *Tipularia* groupings are composed of a series of many corms in different stages of maturity. The parent corm is attached by rootlets to the daughter corm (fig. 10.2). Corms persist for several seasons until they lose substance and gradually wither. Daughter corms contain stored food sources for the current growth season.

Corms and fibrous rootlets are extremely shallow; they lie near the surface of the soil. The conditions of rain, snow, and heavy winds can uproot and transfer them to a new area to overwinter. Uprooted corms can be found with the developing leaf still attached. If light and soil conditions are suitable, corms will initiate growth and flourish at the new location. Frequently the corms are carried into ditches, streams, and sewage drains. Corms perish in this manner.

Natural losses of corms are due to the white-tailed deer (*Odocoileus virginianus*; fig. 10.3) and the white-footed mouse (*Peromyscus leucopus*).

These animals are known to forage on corms throughout the season, especially when other food sources of vegetation are reduced.

In recent years, however, losses have been impacted by an increase in human activities. The author has documented the eradication of numerous established colonies by ongoing construction projects in the Nashville area.

Foliage

During late September, a solitary funnel-shaped leaf emerges from the humus. Standing vertically, it has a translucent sheath wrapped around the inward curving leaf margins. It ranges in color from dusty mauve to brown. It is borne on a sturdy petiole ranging from 2.5–5.5 cm long. In plants located within dense, closed canopies, petioles may be as long as 12 cm as they extend toward the sunlight.

FIGURE 10.4. These *Tipularia* leaves have burgundy markings on a dark green background.

FIGURE 10.5. The bright purple abaxial side of *Tipularia discolor* leaves.

As leaf borders start to unfurl, the maturing leaf becomes ovate in shape (fig. 10.4). Leaves range from 4.5–11 cm long and 3–6.5 cm wide.

Pigments of chlorophylls, anthocyanins, and xanthophylls produce variations of color within the cells of the leaves.[4] The adaxial (upper) surface of the leaf displays variation in both color and pattern. In mature leaves, the background color ranges from a vibrant green to a dusty, dark green. Patterns range from deep red, purple, or frosted spots that are scattered to the leaf margins. Some leaves are solid green with a distinct corrugation. No matter the color variation of the adaxial side, the abaxial (lower) surface is a uniform, vibrant shade of purple (see fig. 10.5).

Tipularia discolor is considered a wintergreen orchid. They are highly adapted to photosynthesis at varying degrees of light levels while sustaining lower climate temperatures.[5]

Leaves begin senescence and decomposition from April into May. No leaves are present at the time of flowering.

The Flowering Cycle

A curved inflorescence, in the shape of a tiny dragon head, emerges in late July (fig. 10.6). It contains tightly packed immature buds that rise out of a translucent red-yellow sheath. As the dark red buds mature, they will have ample space between them to allow for full development of each flower. Buds open in an ascending manner.

The mature inflorescence ranges from 9–28 cm in height, with the total plant (peduncle and inflorescence) reaching upward to 65 cm tall. Sixteen to fifty-five pendant-like flowers may be produced (see fig. 10.7).

These orchids are camouflaged by their colors and blend in with the surrounding vegetation. Three oblong sepals and two petals are positioned in a modified orientation. They are described as left-handed and right-handed flowers.[6] Sepals and petals have a green background

FIGURE 10.6. The developing inflorescence appears as a tiny dragon head rising from the earth complete with a sheath and abundant flower buds.

FIGURE 10.7. This impressively tall *Tipularia discolor* inflorescence measured 63.5 cm. Note the fern frond that surrounds the orchid.

with dark red to purple minute veins running through them. Sepals are 4.5–8 mm long and 1.5–2.5 mm wide, while the petals range smaller in length, 4–7 mm and in width, 1–1.7 mm.

The cream-colored labellum is triple lobed, consisting of two basal lobes and one narrow central lobe that terminates in an upward, flared curve. It measures 4.5–7.5 mm long and 2.5–3 mm wide. The column appears glowing green when observed at nautical twilight. It measures 2.5–4.5 mm in length. Two pairs of pollinia, ovate in shape, are located in the column and attached to a viscidium.

A nectar-filled spur originates from the base of the labellum. The color of the spur is a translucent green that extends behind the flower. It ranges from 8–23 mm in length. The meniscus of nectar in the spur is easily detected by the unaided eye.

Pollination Mechanisms

The unique pollination mechanism in *Tipularia discolor* is due to the asymmetrical shape of the flower. The design of this modified symmetry allows for correct placement of the pollinia, ensuring productive pollination (fig. 10.8). Pollinia become attached to either the left or right side of the compound eye of the insect. When a potential pollinator forages from flower to flower, it transfers the pollinia onto the glutinous stigma surface.

During the daylight hours when temperatures range from 75-85° F, the orchids host a flurry of activity from a variety of insects. Small bees such as the metallic *Augochlora pura* are frequently seen foraging the blossoms. Nurseryweb spiders such as the *Pisaurina mira* cleverly build webs between the individual flowers (see fig. 10.9). Flies in the

FIGURE 10.8. The microscopic view of the pollinia of *Tipularia discolor*.

FIGURE 10.9. A nurseryweb spider, *Pisaurina mira*, on an inflorescence of a *Tipularia* is concealed in her web.

FIGURE 10.10. A noctuid moth searches for nectar rewards at astronomical twilight.

Diptera order zoom in and out among the flowers and provide spiders with a quick lunch. Because fragrance plays a large part in attracting pollinators, fertilization is uncertain as the scent of the orchid at this time of day is muted.

Geometrid and noctuid moths are the principal pollinators of the orchid in the evening hours. The fragrance of the orchid is detectable at nautical twilight and increases in intensity as the night progresses. This potency in scent correlates with the timing of the moth's appearance. The noctuid moth *Pseudaletia unipuncta* and the geometrid moth *Protoboarmia porcelaria* are prevalent foragers, along with a variety of other small- to medium-size noctuid moths (fig. 10.10).

The fragrance of this orchid is pleasant, fresh, and airy. It can be compared to the herb lemon balm or citronella grass. Using scented products on the body or clothing can hamper the ability to detect these incredible orchid fragrances. (See "A Special Note on Detecting the Fragrances of Native Orchids" in Appendix 1 for more information.)

Seed Set, Seed Dispersal, and Wintering

After pollination, mature seed capsules are apparent in early autumn. The young seed capsules, 9-12 mm long, are green with purple shading. When mature, they become brown and desiccated (fig. 10.11). The capsules have five to six seams that split vertically to release a vast number of seeds. With the unaided eye, seeds appear as fine dust (fig. 10.12).

Seeds fall to the ground within the area of the colony. They become scattered among the rootlets of the parent plants. Mechanisms that aid seed dispersal are movements of air currents, rain, and drifts of snow.

Split seed capsules look like tiny skeletons that shake and rattle in the wind. The author noticed during the winter season that the desiccated hull of the seed capsule held an abundant source of rainwater. Several types of insects used their proboscis to draw up water from the capsules, providing refreshing hydration.

FIGURE 10.11. Seed pods appear after the flowers have been successfully pollinated.

FIGURE 10.12. A comical view of a *Tipularia* seed as it goes for the microscope pointer!

FIGURE 10.13. This cheerful spirit's furry coat carried and distributed *Tipularia discolor* seed samples.

A unique and surprising method of seed dispersal is through native animals. Seeds can get lodged in animal fur as they forage through the woods. The author took samples from individual hairs of a young raccoon (*Procyon lotor*; fig. 10.13). The hairs, seen under a microscope, retained substantial amounts of *Tipularia* seeds.

The wintering season of *Tipularia discolor* is in contrast to many wildflowers and woodland plants. Colonies or solitary individual orchids exhibit fresh green foliage that persists throughout the colder months, while most other woodland plants are in senescence.

Observations on Special Characteristics

AN UNUSUAL, COMPACTED FLOWER

A small, densely compacted *Tipularia discolor* was discovered among an established colony of seven uniformly sized plants on August 12, 2021 (see fig. 10.14). This orchid measured 25.4 cm tall in comparison to the other individuals in the colony, which ranged from 42–44 cm in height. This variant was not observed in the preceding years of 2015 through 2020, nor did it appear in the subsequent years of 2022 and 2023.

This individual had densely compacted flowers, especially toward the apex. Flowers were pale in color. Light bluish-purple veins ran through the petals and sepals. The column along with the labellum was solid white in color. It was hypothesized by the author to be a cerulean variable. The orchid was noted to have conventional floral parts, including intact pollinia. Although it was foraged by various insects, it produced no seed capsules in the autumn. Aesthetically, it was conspicuous.

FIGURE 10.14. This unusual, compact *Tipularia* was discovered among standard-sized plants.

FIGURE 10.15. The orchid's name originated from this insect, the crane fly.

THE RELATIONSHIP WITH THE CRANE FLY

The crane fly has long been associated with *Tipularia discolor* (fig. 10.15). The arrangement of the flowers around the inflorescence are similar in appearance to the flies in flight. In North America, there are approximately 1,500 species of crane flies.[7] Larvae live in an aquatic environment in moist soil and wetlands. Mature flies inhabit bogs and swamps. They serve as important food sources for many species of fish and other aquatic wildlife. Unlike mosquitoes, crane flies do not bite.

FOLKLORE OF *Tipularia discolor*

MY WORLD WAR I FRIEND

In a tiny remote area of West Tennessee, the violent earthquakes of 1811 and 1812 formed Reelfoot Lake. Native American legend relates a tale of the Great Spirit who showed his anger by causing the waters of the Mississippi to run backward, creating the crescent-shaped waters of the lake. It is a lovely, mysterious place with an abundance of unique animals and plants.

The greening of agriculture is picturesque, with immense fields of corn, cotton, soybeans, and wheat. Magnificent colors of the sunset can be seen by twilight, while glowing constellations are unobscured by bright city lights.

In the early 1960s, my family had a friend who was a World War I veteran. He and his pal spent their days in a tiny, corrugated, tin-roofed hut on the banks of the Mississippi River. Each day they fished with worn cane poles to catch their meals. They cooked on an open fire. These men had very little in the way of material comforts. I never questioned whether they lived this way by choice or whether they had limited options.

My parents showed great compassion for their welfare. They bought them bunk beds to fit into their dirt-floor abode, books to read, and gifts of prepared food. We would visit them in the late afternoons when my father got home from his work.

For enjoyment, my friend would read the dictionary every night. He developed an extensive vocabulary. With this skill, coupled with a creative imagination, he had a remarkable gift for storytelling. When we visited, he often told entertaining tales to my family.

Once he narrated a mysterious tale about a "crane-fly plant." This plant had magical powers and would turn into a will-o'-the-wisp at the

FIGURE 10.16. The green glow of *Tipularias* at night.

stroke of midnight. He would begin his tale in a soft, low voice:

"Did your daddy show you that plant over there in the woods? Well, it really ain't a plant at all; it's a will-o'-wisp. Did you know what that ole will-o'-wisp did?" he said in a whisper. "It lights the way for the ghost of a Civil War soldier named Steely-Waller. It would be a' lookin' just like a dim candle lit out over the moonless harvest field over there. It would sputter and go out now and then, but it would keep coming back.

"Now this soldier got home after the war. He was searching for his sweetheart, a girl by the name of Lillianette. He wanted to bring her some flowers but most of them were all gone, except the crane-fly ones. He picked as many as he could one late evenin'. Little did he know she had died a year before from the fever."

My friend paused, took a long sigh, and continued, "She was a beaut, but he didn't get to see her. He roamed that field for no tellin' how long a time. The town folks told him that she was buried in that field, but ole Steely-Waller didn't want to know it. He didn't even want to hear about it. They said he finally died of a broken heart. To this day he's still a' roamin' and that crane fly is still a' burnin'. So, if you ever see this crane fly at sundown, be careful and run as fast as you can, 'cause the ghost of that ole soldier is close by. He might just run after you and give you nightmares," he said with a laugh.

After the story ended, we shivered in delight. This sweet story has remained with me throughout my life.

FIGURE 11.1. The noble *Cypripedium parviflorum* var. *pubescens*.

11

Cypripedium parviflorum var. *pubescens*

The Shoe of Venus

ONE OF THE MOST INTRIGUING GROUPS of orchids in the plant kingdom is the genus *Cypripedium*. With their slipper-like flowers in brilliant tones of color, colonies seen in the wild can be astonishing to the eye (fig. 11.1).

Common names for members of this genus are the Shoe of Venus, Moccasin Flowers, Whippoorwill Shoes, and Greater Yellow Lady's Slipper. For this guide, the orchid featured will be referred to by its synonym, *Cypripedium pubescens* (Willdenow) O. W. Knight.[1]

The orchid spans from Alaska across to Canada, from eastern North Dakota down to New Mexico, and throughout the eastern and southern states, excluding Louisiana and Florida. They have been found in a variety of habitats ranging from forested bogs, glades, mesic coniferous forests, and open-canopy deciduous woodlands. Rarely found in Davidson County, these orchids grow in sandy loams or the slightly acidic humus-enriched soils under deciduous trees.

There are many threats to this species due to the orchid's dramatic appearance. Expansive transplantation from their natural habitats has caused a reduction in the number of colonies.[2] These wild plants are frequently sold in commercial markets for use in home gardens.

Historically, the roots of the *Cypripedium* species have been collected for use in folk medicines.[3] In the United States, the harvesting of orchid roots is discouraged. Sustainable native plant nurseries have commercially propagated plants for retail sale. This current method helps to reduce unethical wild-collecting and harvesting. Agar-generated seed germination and tissue culture are future options for mass cultivation. This will serve as a valuable aid for the preservation of the species.

HELPFUL HINT

Cypripedium pubescens is sparsely distributed in the Central Basin. A limited number of plants can be observed in parks, greenways, and private gardens. Whether these plants are naturally occurring or have been relocated from other areas is not known. Additional locations in Tennessee for observing members of the *Cypripedium* species include the Cumberland Plateau and the Great Smoky Mountains National Park.

Cypripedium pubescens blooms in the months of April and May in Nashville.

Rhizomes

Underground rhizomes are thick and fibrous. They are produced by the plant annually and grow from the anterior end, as the posterior tip diminishes.[4] Branching out from the rhizomes are yellow to brown roots with a tough, fibrous consistency. Unearthed, they look like a tangled mass of spaghetti noodles (fig. 11.2). It requires several years of growth to produce a large root mass.

FIGURE 11.2. Roots of a *Cypripedium* species with new shoots of growth appearing at the top of the root.

FIGURE 11.3. New leaves emerging with a crinkled and corrugated appearance at the top of each leaf.

Rhizomes produce clonal offsets from the parent plant. Several shoots can arise from a single rhizome in the spring. This is often the case when many young plants are found occurring in dense colonies. They overwinter after the current season's growth cycle, protected by the soil and humus.

Foliage

In April, a bright green shoot arises from the soil. A sheathing bract encircles three to five young, pubescent leaves. As the leaves spread horizontally, they appear crinkled and folded (fig. 11.3). They have a corrugated appearance due to the pleating of veins that run parallel to the leaf margins. Leaves continue to be covered with white pubescent hairs throughout their life cycle.

The leaves are alternate and evenly spaced on the stem with smooth margins. Mature leaves are elliptic-lanceolate or broadly ovate in shape, tapering at the end. Leaves range in measurements of 8–20 cm long and 5–10 cm wide. The central stem terminates in a flowering bud.

This orchid has an abundance of dense glandular hairs located not only on the leaves and stem, but on the floral parts of the petals and sepals as well. The labellum, however, is smooth. It is important to note that when touched the glandular hairs may cause contact dermatitis in sensitive individuals.

The Flowering Cycle

In April, warm temperatures and spring rains bring about the flowering season for *Cypripedium pubescens* in the Central Basin. After the orchid has reached a height ranging from 12–77 cm tall, buds begin to swell at the terminal ends. Bright green sepals and petals surround the concealed labellum (see fig. 11.4).

Note that this species is highly variable. In certain soils and climates, flora and foliage can have significant differences in measurements. The measurements here reflect the plants located in the Nashville vicinity.

FIGURE 11.4. Green leaves and other floral parts conceal a hidden yellow labellum. Notice the large leaf-like bract above the terminal bud.

FIGURE 11.5. Much like a bright yellow Easter egg, the oval labellum makes an entrance. The dorsal sepal and the immature petals can be seen. Petals become twisted with maturation.

The dorsal sepal is ovate to lance-ovate in shape and covers the top of the labellum. It measures 4–7 cm long and 2–3 cm wide. The coloration of the sepal is greenish-yellow with faint streaks of rusty-brown.

The two lateral sepals are 3.5–7 cm long and 1–2.5 cm wide. Fused behind the labellum, the lateral sepals are termed *synsepalum*, which describes the fusion of two or more sepals creating a compound structure.[5]

Spiraling petals add to the allure of this orchid. The two petals are linear-lanceolate to lance-ovate in shape, 4.5–9 cm long and 0.3–1 cm wide. The petals and sepals vary in color from yellow-green to yellow-brown. They have linear spots of dark red to brown dispersed throughout the length.

A solitary leaflike bract stands erect behind the ovary. With maturation, this bright green bract measures 4.5–9 cm long and 2–4 cm wide. It is often mistaken for the dorsal sepal.

As the floral parts start to unfold, a bright yellow labellum becomes apparent (fig. 11.5). The labellum defines the characteristic slipper-shape of the orchid. It is ovate in shape, and on an average approximately 4.5 cm long with inward rolled edges at maturity. This floral morphology is significant to the unique manner in which this orchid carries out pollination.

Pollination Mechanisms

The complex flowers of the *Cypripedium* species are generally self-compatible; however, most of the flowers are pollinated by various types of bees.[6]

Cypripedium pubescens orchids are visited by a diverse group of insects that includes flies (Diptera), bees, and wasps (Hymenoptera). Successful pollination is determined by the insect's size and behavior characteristics. The author observed various bee types: bumblebees (*Bombus*), small carpenter bees (*Ceratina*), metallic-colored sweat bees (*Augochlora pura*), and mason bees (*Osmia*).

FIGURE 11.6. An *Osmia* bee blown into the labellum by the wind.

Attracted by the slipper's bright color, potential pollinators land on the top of the labellum or on the staminode. Inadvertently, they slip on the inward rolled edges and fall into the depth of the labellum (fig. 11.6). The author observed many types of unsuspecting insects blown into the depth of the slipper by the wind. The labellum's glossy, smooth margins prevent any effort from the insect escaping the way it entered.

After many unsuccessful attempts to exit, the insect is guided by hairs and light windows to a vertical passageway. The light windows are translucent, tissue-thin areas located underneath the adjacent side openings (exit holes) of the labellum. They have been presumed to direct the insect to crawl in the correct direction toward an exit (fig. 11.7).[7]

If the insect is of the correct size and shape, it will be able to squeeze under the stigmatic surface. Next, it will squeeze through one of the two lateral anthers. A glutinous mass of pollinia partially blocks the exit holes. The insect exerts pressure by squeezing through the hole, which allows the pollinia from the anther to adhere to either their head, eyes, thorax, abdomen, or wings. The insect, with the pollinia, may enter another labellum and repeat this exit process. In this manner, the pollinia is transferred to a different stigmatic surface or reintroduced into the same flower.

The author noted individual *Cypripedium pubescens* orchids had subtle fragrance differences. Most were similar to a rose-like musk, while a few possessed a slightly warm scent, similar to an oven-baked dessert. The fragrance mimics pheromones to attract insects for possible nectar rewards.[8] However, this orchid is deceptive and offers no such rewards.[9]

Seed Set, Seed Dispersal, and Wintering

After successful pollination, a three-chambered ovary, located behind the labellum, starts to swell. The green, expanding capsule measures approximately 2.5–3.6 cm at maturity. By October, it becomes brown, desiccated, and sharply ribbed. Remnants of the dried flower can be seen attached to the apex of the capsule (see fig. 11.8).

Although most orchid seeds are extremely small, ranging from 0.05 to 6 mm, *Cypripedium pubescens* seeds are large in comparison, 1.19 mm in length (see fig. 11.9).[10]

Cypripedium pubescens seeds are dispersed by rain, snow, and wind currents. The seeds lack a food source and must depend on the mycorrhizal fungi association for successful development of the embryo. Seeds, stored correctly with the proper conditions, remain viable for many years.[11]

FIGURE 11.7. A *Ceratina* bee finally discovers an exit.

Observations on Special Characteristics

THE STAMINODE

Located above the opening to the labellum is the triangular to ovate-shaped staminode. This striking floral part of the orchid appears much like a flexible shield. It is a modified sterile anther with blunted edges. It is decorated with maroon or red dots, which may serve as a lure for pollinating insects (fig. 11.10).[12] The abaxial surface has a raised central margin, which also may contain colorful red dots.

The staminode blocks the opening at the base of the labellum, directing insects to the correct exit openings. The trilobed stigma lies vertically behind the column and the staminode. Two fertile anthers are positioned on either side of the column. They appear like eyes looking out through the lateral exit openings (see fig. 11.11).

Foraging insects cannot retrace their entrance into the labellum. They are directed by numerous guide hairs to exit behind the staminode and through the exit openings. Pollination may occur in this manner by deceit.[13]

FIGURE 11.8. Remnants of the dried flower can be seen attached to the apex of this *Cypripedium* maturing seed pod.

FIGURE 11.9. *Cypripedium parviflorum* var. *pubescens* seeds making a V for victory.

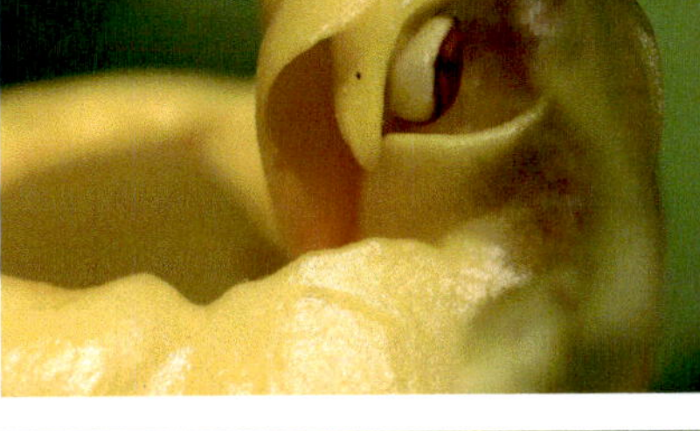

FIGURE 11.10. The staminode contains spots that may be designed to ensure insect pollination.

FIGURE 11.11. Eye-like sockets behind the staminode contain the stigma and style.

FIGURE 11.12. A beetle is directed toward the exit by the tiny stiff hairs within the labellum.

INSECT VARIETIES OBSERVED INSIDE THE LABELLUM

On bright sunny days, there are many types of various foraging insects observed inside the labellum. Bees, flies, and wasps frequent the flowers for nectar rewards. No matter what type of insect gets caught inside the labellum, they are directed under the anthers and must exit through the lateral openings.

The Asian ladybeetle or harlequin beetle, *Harmonia axyridis*, often explores the deep inside of the labellum, which results in the beetle getting trapped (fig. 11.12). Like many other foragers of this orchid, they manage to travel along the guide hairs and find an exit.

Spiders are attracted to the slippers for shelter. Stealthy spiders build webs directly behind the labellum. They are successful in catching a various array of tasty, succulent prey!

FOLKLORE OF

Cypripedium parviflorum var. *pubescens*

For many years in America, *Cypripedium* roots were harvested and administered to children for their calming effects. The Lady's Slipper roots contain hypnotic, nervine, and sedative medicinal properties. Traditional Chinese medicine has utilized these orchids for thousands of years.

MRS. L. S. LOUIS

Mrs. L. S. Louis, by today's standards, would be considered a professional nanny. She loved children and many of the children she helped raise regarded her as a prominent mother-figure.

Mrs. Louis was often the first person a child saw in the mornings when they woke up. As she prepared their breakfast, she would tell them wonderful stories. Her stories varied in subject matter, but mostly the stories were about her family in days long ago.

Sunday was her favorite day. After church, she would go with her husband to the state woods to collect a variety of natural foods. Paw-paw fruits, persimmons, hickory nuts, and walnuts were collected and made into delicious breads and cakes. She would gather grapevines in the late autumn to make into seasonal wreaths. She would sell these to local folks in town to supplement her household budget. Her wreaths were displayed on many doors and over fireplaces at Christmas.

In the spring, she collected herbs for soothing teas and to stock her family's medical chest. To decorate her home, she knew which wildflowers would hold their color for dried arrangements. She enjoyed talking about the medicinal properties of these plants.

One day, her story was a tale about little Herbert, a child who would never go to sleep at night. She said her grandmother told this story to her when she was very young.

FIGURE 11.13. A *Bombus* species searches for nectar rewards.

She began her story, “Now Herbert was a boy with troubles. He had the colic and the sore throat and he never slept. So Grandma Dot got ’round in the woods and dug some slippers. She had to dig a whole lot ’cause the roots were stout and tangled up in the ground.” She added, “Slippers have something in ’em that would help little Herbert sleep.

“Grandma Dot would mash and then boil those roots down. She put them into a big crock with some sugar and some other ingredients. This was set in the root cellar for a long time, but I don’t remember how long it took. It did set little Herbert right.”

She continued, “When he woke up, he was calmed down. We didn’t have to dose him up with high-cost medicine from the store. He didn’t mind the taste of the slippers, but he would spit out the doctorin’ medicine every time! He grew up to be a big strong man, but that was a long time ago,” she laughed.

Mrs. Louis was quite a character. I will always remember some of the stories she related in various interviews. She sent me a Christmas card the year before she passed away. I have kept this treasured card since.

When we devalue wildlife in all its forms, we devalue human life as well.

EPILOGUE

The Ultimate Gift

ORCHIDS REPRESENT some of the most highly evolved plants in the world. A single native orchid holds a universe all its own. Among its corms, leaves, and flowers are a galaxy of living things. Underneath the soil, roots have an association with mycorrhizal fungi for growth and development. This microscopic world is flourishing with complex biological actions.

Orchids are among some of the first plants to become threatened or endangered.[1] The plants and the numerous life-forms that they support are highly susceptible to slight changes in the environment. When a single orchid is destroyed, numerous microscopic life cycles are also halted.

Pre-Columbian Indigenous societies had a working knowledge of local animals and plants. They expressed reverence for all the natural resources that supplied their needs. Today, *Homo sapiens* depends less upon the natural world. We may approach these impinging questions: Should we assume that there are unlimited resources available for future use? Is nature only a device designed to provide humans with opportunities to benefit their needs? The answers depend on personal ethics.

How can we meet these challenges locally? To get started we can conduct our own research:

1. Get out your walking shoes and take the family on a hike, not only for recreation, but to get acclimatized to the surrounding world of nature.
2. Bring some guidebooks along. Make a list of the flowers and trees that you see. Become familiar with the common names of the plants and trees you observe.
3. Look for any signs of animal life. Make notes and draw pictures of their footprints, if possible.
4. Study varied subjects such as climate change, pollination, and wildlife habitats.
5. Look for native orchids in area parks and greenways. Observe their pollinators.

After following these suggestions, you may notice that you will become more attentive to your carbon footprint. You may start to make changes in the way your handle your garbage, plastic usage, and chemicals. Becoming knowledgeable about your surroundings is the first positive step one can take to make the change needed in your neighborhood.

When we devalue wildlife in all its forms, we devalue human life as well. It all depends on what we wish to conserve.

APPENDIX 1

LOCATING NASHVILLE NATIVE ORCHIDS THROUGHOUT THE YEAR

Taking an excursion to locate terrestrial orchids can be exciting. Cataloging the plants with a drawing tablet or a camera adds to the enjoyment.

Here are some tips:

1. Vividly colored *Galearis spectabilis* is a tiny orchid that is found in semi-closed canopies in the month of April.
2. *Goodyera pubescens* orchids have prominent, eye-catching leaves that make it easy to locate throughout the seasons.
3. *Liparis liliifolia* is brightly colored. Look for the characteristic large purple labellum in mid-May.
4. The *Spiranthes* species of orchids can be found within the perimeters of lawn boundaries in the late summer and early autumn.
5. *Tipularia discolor* has a conspicuous solitary leaf that predominates in the winter season. Marking the winter location with natural objects such as stones makes it easier to identify the camouflaged bloom during the late summer.
6. The slipper orchid, *Cypripedium parviflorum* var. *pubescens*, has a bright colorful lip that can be seen from far away. It stands out like a beacon in April and May.

A Special Note on Detecting the Fragrances of Native Orchids

Native orchid fragrances are varied. Many are fresh and exhilarating, while some would only cater to a fly! The intensity depends on the temperature and time of day or night.

The author has noticed that regular use of scented products, especially air fresheners, body and home deodorants, laundry detergents, and potpourris dull the sensitive olfactory senses, making orchid fragrances difficult to distinguish. Wildlife, including insects, can detect synthetic fragrances from great distances; they leave the area quickly. This can prevent a person from obtaining remarkable photography shots in the wild.

APPENDIX 2

A CONSERVATION NOTE

There are many ways to get involved with orchids on a conservation and preservation level. Here are some suggestions to help with these efforts for the native orchids in your area:

- Terrestrial orchids should never be collected or dug out of their natural habitat. Infringing on the regulations of protective conservation plus trespassing on private property are serious matters. Always obtain written permission from the landowner before exploring new property.

- Orchid rescues should only be conducted under the supervision of an experienced person who has legal permission to secure and transfer the plants on commercial or private property. Make sure the site is safe before starting a rescue. Have a balanced plan to coordinate the time allotment for plant rescues with the landowner, construction supervisor, or timber manager. If volunteers are involved, have a detailed list on how the rescue will be conducted and print out copies for them. Plan to leave the place as clean and untracked as possible. Park in designated areas only.

- If you wish to cultivate and enjoy growing orchids, research information about tropical hybrids. Many can be grown on a windowsill in your home. Healthy plants are readily available from local orchid shows and sustainable growers in your area.

- Join a garden club to collaborate with other growers in your community. Consider a membership with an orchid alliance such as the American Orchid Society and the Orchid Digest. These organizations have information and announcements about orchid culture and conservation, plus a list of current locations for orchid shows.

APPENDIX 3

AN EXPLANATION OF SCIENTIFIC NAMES

A scientific name consists of two Latin words, genus and species. These words are used by scientists and plant enthusiasts all over the world. For example, with the terrestrial orchid *Tipularia discolor*, *Tipularia* is the genus and *discolor* is the species name.

So what are the abbreviations and proper names listed after the genus/species names? The abbreviation (L.) refers to Carl Linnaeus. He was a Swedish scientist who is credited with establishing the system of modern binomial nomenclature, the identification of plants by a genus followed by a species name.[1]

Besides Linnaeus, other names listed are names are the person(s) who described and published the name of the orchid. An example is *Willdenow*, often located in texts after the genus/species name *Goodyera pubescens*. This person gave validity to the plant name, which is considered the prescribed name for use.

Genus names are often changed through a process of reclassification. Today, with the advent of molecular genetic testing, orchids are reclassified on the basis of their genetic code, their DNA analysis. This is especially true for orchids that grow in tropical environments.

The rapid changing of names can be confusing for the budding orchidist as well as the general horticulturist. For example, *Brassavola digbyana* is now named *Rhyncholaelia digbyana*. It helps to learn a few names at a time, instead of tackling a host of names all at once.

APPENDIX 4

AN EXPLANATION ON THE FOLKLORE SECTIONS

When I was a young child, I wanted to interview people like I heard on the radio. My parents told me to take careful notes with pen and paper as I asked questions. I followed their directions. Years later, I asked my parents for a cassette player with tapes for my birthday present. I was very happy when I opened the bulky package and saw the player plus several cellophane-wrapped tapes. Over the years, I took notes and recorded anyone who would allow me the time and had the patience to sit for my interviews. I went to private residences and nursing homes (as they were called in those days) and taped many people who had been acquainted with my ancestors. Although these people passed on years ago, they left a legacy with me via notes and cassette tapes. Little did I realize that forty to fifty years later I would be using some of their interviews in this guide for the Folklore Sections.

Some of the names have been omitted or changed due to consideration for the remaining families' privacy.

The mode of speech is presented here exactly as it was related in its various forms for historical content only.

IMPORTANT NOTE

The stories do not dispense medical advice or prescribe the use of orchids and other plants as any form of treatment. It is not the intent of this guide to diagnose or prescribe. The information presented here is of historical interest only.

ACKNOWLEDGMENTS

Grateful acknowledgment to LinnAnn Welch for composing the Foreword.

Special thanks to Tavia Cathcart Brown for enthusiastic encouragement; to J. Edgar Lowe for allowing "free-range photography"; to Colton Cates Dunstan Parr for hours of technical aid and emotional support; to David Harwell Parr for lenses, lights, navigational guides, knee pads, and patience; to the team at Vanderbilt University Press and the editorial committee for believing in the project.

A hearty thanks to the following people for advice, information, and land access (in alphabetical order):

Krista Allen, Jenny Andrews, Dr. Joseph Arditti, Robert Buck, Ron Buck, Arthur Everett Chadwick, Dr. Frank A. Hale, Tom and Laura Harper, Dennis Horn, Tom King, Randall Lantz, Kay McDonald, Julianne McGuinness, Dr. Ron McHatton, Dr. Steve Murphree, Dr. Katie Owens-Murphy, Dr. Matthew Pace, Richard Page, Milo Pyne, Dr. Sarah Rose, Dr. Laura Russo, Dr. Jack Santino, Dr. William C. Schumann, Dr. Louis N. Sorkin, Rita Venable, and Lou and Tom Walker.

GLOSSARY

abaxial	the outer or under surface of a leaf or floral part directed toward the main axis
achlorophyllous	a plant lacking chlorophyll, usually not photosynthetic
adaxial	the inner or upper surface of a leaf or floral part directed toward the main axis
alternate	leaves arising singly at a node, not in pairs from the stem
anther	a portion of a stamen (male part) that contains pollinia grains
anthesis	the opening of the functional flower; the opening of the stamen
apex	the tip end of a part of a plant
arthropods	invertebrates such as insects, arachnids, and crustaceans
asexual pollination	fertilization that is independent of a sexual process
astronomical twilight	the period after sunset ending when the sun is between 12 and 18 degrees below the horizon
autogamous	self-pollinating without the aid of a pollinating agent

axil	an angle between the stem and upper side of a leaf
basal	arising from the base of a stem
binomial	a two-part name, consisting of the genus followed by species
blade	the part of a leaf that is flat and broadened
bract	a leaf-type structure that is modified at a node on or below the flower
Calvin cycle	the dark reaction of photosynthesis that does not require light
capsule	the dehiscent part of the fruit (a ripened ovule and its parts) that holds seeds
carbohydrates	compounds consisting of carbon, hydrogen, and water that produce energy by oxidation; these include cellulose, sugars, and starches
carbon footprint	the amount of carbon dioxide emitted by something, such as a person's activities or a product's manufacture and transport
chlorophyll	a green pigment in plants that traps energy from the sun to begin the process of photosynthesis; it contains a tetrapyrrole ring with a magnesium center
civil twilight	the period after sunset ending when the sun is at most 6 degrees below the horizon
column	in orchids, a modified part formed by the pistils and stamens
cordate	having the shape of an ace of spades or heart-shaped
corm	an underground storage organ; an enlarged bulb
corolla	a collective name for petals, sepals, and other parts of a flower

cotyledon	a leaf that emerges from inside the seed of a plant; it is usually different in shape than the mature leaf
crenate	scalloped in shape
deciduous	trees, shrubs, and foliage that drop leaves at a particular cyclic season
dicotyledon	a double seed-leaf, or cotyledon, within the seed of a plant
disk	the top surface of the labellum, or lip, in an orchid flower
distal	a part farthest away from the point of attachment to the main body
dorsal	the outer or upper side
ellipsoid	oblong with the ends rounded, as in a specific leaf shape
embryo	the small plant that holds the genetic material inside a seed
endosperm	the food tissue that surrounds and nourishes the embryo of a seed
epiphyte	a non-parasitic plant that grows attached to another plant or surface
fungi	the plural of fungus
fungus	a range of dissimilar parasitic or saprophytic organisms that lack chlorophyll including mushrooms, slime molds, and yeasts
fused	parts of a plant body joined together
genera	the plural of genus
genotype	the genetic information about an organism
genus	a group of related plants with distinct characteristics

glabrous	lacking pubescent hairs; smooth
glandular	secreting structures on a plant that hold nectar, oil, or liquid
globose	spherical in shape
grana	sites within the chloroplast where the light reactions of photosynthesis take place
habitat	the natural environment of plants or animals
herbaceous	a plant without a woody stem
herbarium	a collection of plant parts identified and preserved by drying and pressing
hood	a covering located on the apex of a flower
humus	organic matter in the soil that is in a decomposing state
indigenous	native or natural to an area, county, state, or country; not imported
inflorescence	the arrangement of flowers on a stalk or stem
labellum	a modified petal, such as the lip in an orchid flower
lanceolate	spear-shaped; broad at the base and narrow at the apex
linear	narrow and long throughout the length of a plant or organism
lip	a modified petal; another name for labellum
lobe	the indentations at the margins forming a division at or around a flower or leaf
meniscus	in physics, the convex or concave upper surface of a column of liquid
mesic	moist, humid conditions
monocotyledon	a single seed-leaf, or cotyledon, within the seed of a plant

morphology	the structure of a plant or animal
mycorrhiza	an association between plant and fungus that produces benefits for development and growth
nautical twilight	the period after sunset ending when the sun is between 6 and 12 degrees below the horizon
node	the jointed area where leaves are inserted into the stem
oblanceolate	broadest area from middle to apex and tapering at the base
orbicular	rounded or circular
obtuse	blunt or rounded at the apex
opisthosoma	the abdominal region of a spider
ovate	a leaf or flower part that is egg-shaped, with the wider end of the leaf near the base
ovary	the basal portion of the pistil that contains the immature seeds
parasite	a plant that lives on another (plant) body for chlorophyll and nutrients
parthenogenic	able to reproduce without copulation
petal	a portion of the inner part of a flower, typically colored
pedicel	stalks located beneath single flowers that support and attach them to the main axis of an inflorescence
peduncle	the primary flower stalk of a single flower or a cluster of flowers
peloton	fungal structures used by the orchid to support germination
pendent	hanging downward or suspended
petiole	the stalk that supports the leaf

phenotype	the outward appearance of an organism due to the presence of genes
photosynthesis	a chemical reaction in plants that turns carbon dioxide into sugars and oxygen
pinnate	a compound leaf with two rows of leaves on each side of the central stalk
pistil	the female part of the flower that contains the stigma, style, and ovary
pollinia	produced by the anthers; masses of waxy pollen grains; the plural of pollinium
proboscis	any long, flexible snout
proximal	a part nearest the point of attachment to the main body
pseudobulb	pertaining to orchids, a swollen stem for food storage for the plant
pseudocopulation	male insects attracted to a flower, attempt copulation on sexual instincts
pubescent	covered with fine, sticky, or other types of various hairs
raceme	an inflorescence with pedicels that originates on the plant's main axis
reflexed	to be bent, forward or backward
reticulate	net-like in appearance
rhizome	a horizontal underground stem that is distinguished from the roots
rosette	a circular cluster of basal leaves, as in *Goodyera pubescens*
rostellum	a part of the stigmatic lobe that separates the stigmatic surface from the anthers; it produces a sticky gel-like substance for the adherence of pollinia

scape	a flowering stalk or inflorescence, generally without leaves
self-compatible	a species that can be fertilized by its own pollinia and produce seed
self-incompatible	a species that is incapable of effective self-fertilization
sepal	the outer set of elements of the flower that are usually green
sessile	without having a pedicel; without a stalk supporting a flower
sexual pollination	fertilization dependent upon the copulation of two sexes
species	the taxonomic classification in rank after the genus of an organism
spike	any beginning growth or shoot; an inflorescence with sessile flowers
spur	the tubular, hollow extension of the labellum or lip in an orchid flower
stabilimentum	a silk structure included in certain orbweaver spider webs
stamen	the male part of the flower that bears the pollinia
staminode	a modified stamen that is sterile
stigma	the part of the pistil (female part) that receives pollinia
stomata	the pores in leaves and stems through which gas exchange takes place
style	the part of the pistil usually connecting the stigma to the ovary
succulent	the texture of a hydrated, fleshy plant
synsacrum	an elongated sacrum containing a number of fused vertebrae

synsepalum	a fusion of two or more sepals creating a structure
transpiration	the process of foliage losing quantities of water vapor to the atmosphere
taxonomy	a classification system of plants and animals in a systematic order
terrestrial	growing in or on the soil; not growing in the air as an epiphyte, or as an aquatic plant
tuber	an enlarged underground stem containing food storage for the plant
tubercle	a small leaf, bud, or wart-like projection
unicellular	a single cell
ventral	on the inner or lower side
viscidium	the sticky pad or gland of the column that the pollinia attaches to
whorl	a ring of leaves, petals, or flowers
xerophyte	a plant that is able to exist and adapt in very dry, dehydrating conditions

NOTES

Introduction

1. H. W. Crew, *History of Nashville, Tennessee, Illustrated* (Nashville, TN: Publishing House of the Methodist Episcopal Church, South, 1890; facsimile ed., Bristol, TN: Charles Elder, Publisher, 1970), 30.

Chapter 1

1. Glen S. Winterringer, *Wild Orchids of Illinois*, Popular Science Series, vol. 6 (Springfield: Illinois State Museum, 1967), 2.
2. Winterringer, *Wild Orchids*, 2.
3. *The Plant World*, World Book Encyclopedia of Science, vol. 5 (Chicago, IL: World Book, 1992), 18–20.
4. Paola Bonfante and Andrea Genre, "Mechanisms Underlying Beneficial Plant–Fungus Interactions in Mycorrhizal Symbiosis," *Nature Communications* 1, no. 1 (July 27, 2010) : article #48. https://doi.org/10.1038/ncomms1046.
5. Melissa McCormick, Dennis Whigham, and John P. O'Neill, "Mycorrhizal Diversity in Photosynthetic Terrestrial Orchids," *New Phytologist* 163 (June 7, 2004): 426–27.
6. Bonfante and Genre, "Mechanisms," 1.
7. Bonfante and Genre, "Mechanisms," 1.
8. Chihiro Miura et al., "*Bletilla striata* (Orchidaceae) Seed Coat Restricts the Invasion of Fungal Hyphae at the Initial Stage of Fungal Colonization," *Plants* 8, no. 8 (August 11, 2019): article # 280, 2.
9. Marc-Andre Selosse, Julita Minasiewicz, and Bernard Boullard, "An Annotated Translation of Noël Bernard's 1899 Article 'On the Germination of *Neottia Nidus-Avis*,'" *Mycorrhiza* 27, no. 6 (August 2017): 612–13.
10. George H. F. Nuttall, "Symbiosis in Animals and Plants," *American Naturalist* 57, no. 652 (1923): 455.
11. L. van der Pijl and Calaway H. Dodson, *Orchid Flowers: Their Pollination and Evolution* (Coral Gables, FL: Fairchild Tropical Gardens and the University of Miami Press, 1966), 124.

12. D. Charlesworth and B. Charlesworth, "Quantitative Genetics in Plants: The Effect of the Breeding System on Genetic Variability," *Evolution* 49, no. 5 (1995): 911–20.
13. Pijl and Dodson, *Orchid Flowers*, 27–31.
14. Pijl and Dodson, preface to *Orchid Flowers*.
15. Jana Jersáková, Steven D. Johnson, and Pavel Kindlmann, "Mechanisms and Evolution of Deceptive Pollination in Orchids," *Biological Reviews* 81, no. 2 (2006): 220.

Chapter 3

1. Fredericka Martin, *Frances Toor's New Guide to Mexico including Lower California*, 8th ed. (New York: Crown Publishers, 1967): 181.
2. Maria Longhena, *Ancient Mexico: The History and Culture of the Maya, Aztecs, and Other Pre-Columbian Peoples* (Vercelli, Italy: Stewart, Tabori & Chang, 1998), 46–48.
3. Carlos Ossenbach, "History of Orchids in Central America Part 1: From Prehispanic Times to the Independence of the New Republics," *Harvard Papers in Botany* 10, no. 2 (Jan. 2009): 197.
4. Martin, *Frances Toor's New Guide*, 287–88.
5. Editors of Time-Life Books, *The Mystical Year*, Mysteries of the Unknown (Alexandria, VA: Time-Life Books, 1992), 117.
6. Dorothy Hargreaves and Bob Hargreaves, *Tropical Blossoms of the Pacific*, vol. 1 (Lahaina, HI: Ross-Hargreaves, 1970), 47.
7. L. J. Lawler, "Useful Orchids," *American Orchid Society Bulletin* 54, no. 6 (1985): 693–94.
8. "Ice Cream Threatens Turkey's Flowers," BBC World News, August 5, 2003, http://news.bbc.co.uk/2/hi/science/nature/3126047.stm.
9. NatureServe Explorer, s.v. "*Cypripedium parviflorum* var. *pubescens*: Large Yellow Lady's-slipper," page last published Feb. 2, 2024, https://explorer.natureserve.org/Taxon/ELEMENT_GLOBAL.2.134081/Cypripedium_parviflorum_var_pubescens.
10. Brooks Eliot Wigginton, *The Foxfire Book: Hog Dressing; Log Cabin Building; Mountain Crafts and Foods; Planting by the Signs; Snake Lore; Hunting Tales; Faith Healing; Moonshining; and Other Affairs of Plain Living* (Garden City, NY: Anchor Press, Doubleday, 1972), 231.
11. William Sansom, *A Book of Christmas* (New York: McGraw Hill, 1968), 78.
12. "William John Swainson – Biography, Facts and Pictures," Famous Scientists, accessed March 11, 2023, https://www.famousscientists.org/william-john-swainson/.

13. Nora Anghelescu et al., "A History of Orchids: A History of Discovery, Lust and Wealth," *Research Gate, Scientific Papers, Series B, Horticulture* 64, no. 1 (July 6, 2020): 525.

Chapter 4

1. Eldon D. Enger and Bradley F. Smith, *Environmental Science: A Study of Interrelationships*, 8th ed. (New York: McGraw-Hill, 2002), 389–90.
2. US EPA, "Climate Change Indicators in the United States," US EPA, Reports and Assessments, November 6, 2015, https://www.epa.gov/climate-indicators, 3.
3. US EPA, 3.
4. US EPA, 12–15.
5. L. C. De, S. P. Vij, and R. P Medhi, "Impact of Climate Change on Productivity of Orchids," Poster presented at 5th Indian Horticulture Congress, PAU, Ludhiana, Punjab, November 6–9, 2012. ResearchGate, Oct. 2019. https://www.researchgate.net/publication/336252989_Impact_of_Climate_Change_on_Productivity_of_Orchids.
6. Lina Zeldovich, "Bees' and Orchids' Pseudo-Romance Broken by Climate Change.," JSTOR Daily, May 8, 2018, https://daily.jstor.org/bees-and-orchids-pseudo-romance-broken-by-climate-change.
7. De, Vij, and Medhi, "Impact of Climate Change."

Chapter 5

1. Eugene P. Odum and Gary W. Barrett, *Fundamentals of Ecology*, 5th ed. (Belmont, CA: Thomson Brooks/Cole Publishers, 2005), xiii.
2. "Speed-Dating the Spiders II: Ghost Spiders," University of Wisconsin-Milwaukee, College of Letters & Science Field Station, *The Buglady* (blog), April 24, 2018, https://uwm.edu/field-station/speed-dating-the-spiders-ii-ghost-spiders.
3. Sarah Rose, *Spiders of North America*, Princeton Field Guides (Princeton, NJ: Princeton University Press, 2022), 169.
4. Rose, *Spiders of North America*, 188–89.
5. Rose, *Spiders of North America*, 284–85.
6. Ingi Agnarsson, "Morphological Phylogeny of Cobweb Spiders and Their Relatives (Araneae, Araneoidea, Theridiidae)," *Zoological Journal of the Linnean Society* 141 no. 4 (August 1, 2004): 472.
7. Rose, *Spiders of North America*, 547–48,
8. Martin Nyffeler and Georg Benz, "Spiders in Natural Pest Control: A Review," *Journal of Applied Entomology* 103, no. 1–5 (1987): 326.

Chapter 6

1. Charles J. Sheviak and Paul M. Catling. "*Galearis spectabilis*," Flora of North America, last edited Nov. 5, 2020, http://floranorthamerica.org/Galearis_spectabilis. See Appendix 3, "An Explanation of Scientific Names," for information about the different elements of plant names.
2. Gregg Dieringer, "The Pollination Ecology of *Orchis spectabilis* L. (Orchidaceae)," *Ohio Journal of Science* 82, no. 5 (December 1982): 224.
3. Joseph Arditti and Abdul Karim Abdul Ghani, "Tansley Review No. 110: Numerical and Physical Properties of Orchid Seeds and Their Biological Implications," *New Phytologist* 145, no. 3 (March 2000): 367.
4. Kenji Suetsugu, Atsushi Kawakita, and Makoto Kato, "Avian Seed Dispersal in a Mycoheterotrophic Orchid *Cyrtosia septentrionalis*," *Nature Plants*, no. 1 (May 5, 2015): article #15052, 2.
5. Roger Buvat, *Plant Cells: An Introduction to Plant Protoplasm* (London: McGraw Hill UK, 1969): 82–83.

Chapter 7

1. Jacquelyn A. Kallunki, "*Goodyera*," Flora of North America, last edited Nov. 5, 2020. http://floranorthamerica.org/Goodyera.
2. John W. Hayden, "Wildflower of the Year 2016 Downy Rattlesnake Plantain (*Goodyera pubescens*)," *Virginia Native Plant Society* (blog), accessed March 10, 2023. https://vnps.org/wildflower-of-the-year-goodyera-pubescens-downy-rattlesnake-plantain.
3. Hayden, "Wildflower of the Year."
4. Anghelescu et al., "History of Orchids," 524.
5. Henry Baldwin, *The Orchids of New England: A Popular Monograph* (New York: John Wiley and Sons, 1884), 64.
6. Arditti and Ghani, "Tansley Review," 368.
7. Todd C. Bukowski and Terry E. Christenson, "Natural History and Copulatory Behavior of the Spiny Orbweaving Spider *Micrathena Gracilis* (Araneae, Araneidae)," *Journal of Arachnology* 25, no. 3 (1997): 307–9.
8. Nyffeler and Benz, "Spiders," 321.
9. Wigginton, *Foxfire Book*, 230.

Chapter 8

1. Christopher Mattrick, "*Liparis Liliifolia* (L.) L. C. Rich. Ex Lindley Lily-Leaved Twayblade: Conservation and Research Plan for New England" (Framingham, MA: New England Plant Conservation Program, 2004), 2. https://www.nativeplanttrust.org/documents/75/Liparis_liliifolia.pdf.

2. Lawrence K. Magrath, "*Liparis liliifolia*," Flora of North America, last edited Nov. 5, 2020. http://floranorthamerica.org/Liparis_liliifolia.
3. Mattrick, "*Liparis Liliifolia*," 2.
4. Mattrick, "*Liparis liliifolia*," 9.
5. Enger and Smith, *Environmental Science*, 327–28.
6. Dennis Whigham and John P. O'Neill, "Dynamics of Flowering and Fruit Production in Two Eastern North American Terrestrial Orchids, *Tipularia Discolor* and *Liparis Liliifolia*," In *Population Ecology of Terrestrial Orchids*, ed. T. C. E. Wells and J. H. Willems, 89–101 (The Hague, The Netherlands: SPB Academic Publishers, 1991), 91.
7. Whigham and O'Neill, "Dynamics of Flowering," 91.
8. D. E. Christensen, "Fly Pollination in the Orchidaceae," in *Orchid Biology: Reviews and Perspectives VI*, ed. J. Arditti, 415–54 (Ithaca, NY: John Wiley and Sons, 1994).
9. Mattrick, "*Liparis Liliifolia*," 3.
10. Mattrick, "*Liparis Liliifolia*," 3.
11. Mattrick, "*Liparis Liliifolia*," 3.

Chapter 9

1. Jack B. Carman, *Wildflowers of Tennessee* (Tullahoma, TN: Highland Rim Press, 2001), 405.
2. Matthew Pace and Kenneth Cameron, "The Systematics of the *Spiranthes cernua* Species Complex (Orchidaceae): Untangling the Gordian Knot," *Systematic Botany* 42, no. 4 (Dec. 2017): 645.
3. "*Spiranthes lacera* var. *gracilis*," North Carolina State Extension Gardener Plant Toolbox, accessed Feb. 5, 2024, https://plants.ces.ncsu.edu/plants/spiranthes-lacera-var-gracilis.
4. Charles J. Sheviak, "A New *Spiranthes* from the Grasslands of Central North America," *Botanical Museum Leaflets, Harvard University* 23, no. 7 (1973): 293.
5. "Keys and Fact Sheets: Ceratina Bees," BioNET-EAFRINET, accessed Feb. 29, 2024, https://keys.lucidcentral.org/keys/v3/eafrinet/bee_genera/key/african_bee_genera/Media/Html_eafrica/Ceratina_bees.htm.
6. Arditti and Ghani, "Tansley Review," 367.
7. D. G. Lloyd, "Variation Strategies of Plants in Heterogenous Environments," *Biological Journal of the Linnean Society* 21, no. 4 (1984): 357–58.
8. John M. Schmidt and Ann E. Antlfinger, "The Level of Agamospermy in a Nebraska Population of *Spiranthes cernua* (Orchidaceae)," *American Journal of Botany* 79, no. 5 (1992): 501.

9. Howell V. Daly, "Biological Studies on *Ceratina dallatorreana*, an Alien Bee in California which Reproduces by Parthenogenesis (Hymenoptera: Apoidea)," *Annals of the Entomological Society of America* 59, no. 6 (1966): 1,138.
10. Carol Siegal, "The Hairy Story of Orchid Trichomes," *Orchid Digest* 86, no. 1 (Jan.–Mar. 2022): 12.
11. Paul Masonick et al., "Molecular Phylogenetics and Biogeography of the Ambush Bugs (Hemiptera: Reduviidae: Phymatinae)," *Molecular Phylogenetics and Evolution*, no. 114 (Sept. 2017): 226.

Chapter 10

1. Charles J. Sheviak and Paul M. Catling, "Tipularia discolor," Flora of North America, last edited Nov. 5, 2020, http://floranorthamerica.org/Tipularia_discolor.
2. W. P. Stoutamire, "Pollination of *Tipularia discolor*, an Orchid with Modified Symmetry," *American Orchid Society Bulletin* 47, no. 2 (1978): 413–15.
3. Thomas E. Hemmerly, *Wildflowers of the Central South* (Nashville, TN: Vanderbilt University Press, 1990), 11.
4. Nicole M. Hughes et al., "Photosynthetic Profiles of Green, Purple, and Spotted-Leaf Morphotypes of *Tipularia discolor* (Orchidaceae)," *Southeastern Naturalist* 18, no. 4 (Dec. 2019): 641.
5. David Tissue et al., "Photosynthesis and Carbon Allocation in *Tipularia discolor* (Orchidaceae), a Wintergreen Understory Herb," *American Journal of Botany* 82, no. 10 (Oct. 1995): 1,249.
6. Stoutamire, "Pollination of *Tipularia discolor*," 413–15.
7. Donald J. Borror and Richard E. White, *A Peterson Field Guide to Insects: America North of Mexico* (Boston, MA: Houghton Mifflin, 1970), 262.

Chapter 11

1. Charles J. Sheviak, "*Cypripedium parviflorum* var. *pubescens*," Flora of North America, last edited Dec. 14, 2022. http://floranorthamerica.org/Cypripedium_parviflorum_var._pubescens.
2. Sheviak, "*Cypripedium parviflorum*."
3. Sheviak, "*Cypripedium parviflorum*."
4. D. E. Mergen, "*Cypripedium parviflorum* Salisb. (Lesser Yellow Lady's Slipper): A Technical Conservation Assessment" (Lakewood, CO: USDA Forest Service, Rocky Mountain Region, 2006), 17.
5. Mergen, "*Cypripedium parviflorum*," 65.

6. Robert W. Pemberton, "Pollination of Slipper Orchids (*Cypripedioideae*)," *Lankesteriana* 13, no. 1–2 (Aug. 2013): 66.
7. Nicole Forrester, "Inside the Trap of a Yellow Lady's Slipper Orchid (*Cypripedium parviflorum* var. *pubescens*): The Effects of 'Light Windows' and Flower Orientation on the Behavior of a Native Bee (*Andrena Macra*)," Paper 421 (Undergraduate honors thesis, College of William & Mary, 2011), 3.
8. Jersáková, Johnson, and Kindlmann, "Mechanisms and Evolution," 223.
9. Pemberton, "Pollination of Slipper Orchids," 66.
10. Arditti and Ghani, "Tansley Review," 384.
11. Matthew L. Carlson and Justin R. Fulkerson, "*Cypripedium Parviflorum* var. *pubescens* (Willd.) O. W. Knight: Conservation Assessment on the Tongass National Forest, Alaska Region" (Anchorage: Alaska Natural Heritage Program, University of Alaska Anchorage, October 2017), 17, http://accs.uaa.alaska.edu/wp_comtent/uploads/Cyripedium parviflorum_assesement_Final.pdf.
12. Pemberton, "Pollination of Slipper Orchids," 68.
13. Pemberton, "Pollination of Slipper Orchids," 66.

Epilogue

1. Michael F. Fay, "Orchid Conservation: How Can We Meet the Challenges in the Twenty-First Century?" *Botanical Studies*, no. 59 (June 5, 2018): article no. 16, 5.

Appendix 3

1. Anghelescu et al., "History of Orchids," 523.

BIBLIOGRAPHY

Agnarsson, Ingi. "Morphological Phylogeny of Cobweb Spiders and Their Relatives (Araneae, Araneoidea, Theridiidae)." *Zoological Journal of the Linnean Society* 141, no. 4 (August 1, 2004): 447–626. https://doi.org/10.1111/j.1096–3642.2004.00120.x.

Anghelescu, Nora, Anne Bygrave, Sorina Petra, Mihaela Georgescu, and Florin Toma. "A History of Orchids: A History of Discovery, Lust and Wealth." *Research Gate, Scientific Papers, Series B, Horticulture* 64, no. 1 (July 6, 2020): 519–30. https://doi.org/doi: 10.13140/RG.2.2.12020.78727.

Arditti, Joseph, and Abdul Karim Abdul Ghani. "Tansley Review No. 110: Numerical and Physical Properties of Orchid Seeds and Their Biological Implications." *New Phytologist* 145, no. 3 (March 2000): 367–421. https://doi.org/
10.1046/j.1469–8137.2000.00587.x.

Baldwin, Henry. *The Orchids of New England: A Popular Monograph*. New York: John Wiley and Sons, 1884.

Bonfante, Paola, and Andrea Genre. "Mechanisms Underlying Beneficial Plant–Fungus Interactions in Mycorrhizal Symbiosis." *Nature Communications* 1, no. 1 (July 27, 2010): article #48. https://doi.org/10.1038/ncomms1046.

Borror, Donald J., and Richard E. White. *A Peterson Field Guide to Insects: America North of Mexico*. Boston, MA: Houghton Mifflin, 1970.

Bukowski, Todd C., and Terry E. Christenson. "Natural History and Copulatory Behavior of the Spiny Orbweaving Spider *Micrathena gracilis* (Araneae, Araneidae)." *Journal of Arachnology* 25, no. 3 (1997): 307–20.

Buvat, Roger. *Plant Cells: An Introduction to Plant Protoplasm*. London: McGraw Hill UK, 1969.

Carlson, Matthew L., and Justin R. Fulkerson. "*Cypripedium parviflorum* var. *pubescens* (Willd.) Knight: Conservation Assessment on the Tongass National Forest, Alaska Region." Anchorage: Alaska Natural Heritage Program, University of Alaska Anchorage, October 2017. http://accs.uaa.alaska.edu/wp_comtent/uploads/*Cyripedium parviflorum* _assesement_Final.pdf.

Carman, Jack B. *Wildflowers of Tennessee*. Tullahoma, TN: Highland Rim Press, 2001.

Charlesworth, D., and B. Charlesworth. "Quantitative Genetics in Plants: The Effect of the Breeding System on Genetic Variability." *Evolution* 49, no. 5 (1995): 911–20. https://doi.org/10.2307/2410413.

Christensen, D. E. "Fly Pollination in the Orchidaceae." In *Orchid Biology: Reviews and Perspectives VI*, edited by J. Arditti, 415–54. Ithaca, NY: John Wiley, 1994.

Crew, H. W. *History of Nashville Tennessee, Illustrated*. Nashville, TN: Publishing House of the Methodist Episcopal Church, South, 1890. Facsimile edition, Bristol, TN: Charles Elder, Publisher, 1970.

Daly, Howell V. "Biological Studies on *Ceratina dallatorreana*, an Alien Bee in California which Reproduces by Parthenogenesis (Hymenoptera: Apoidea)." *Annals of the Entomological Society of America* 59, no. 6 (November 1, 1966): 1,138–54. https://doi.org/10.1093/aesa/59.6.1138.

De, L. C., S. P. Vij, and R. P. Medhi. "Impact of Climate Change on Productivity of Orchids." Poster presented at 5th Indian Horticulture Congress, PAU, Ludhiana, Punjab, November 6–9, 2012. ResearchGate, Oct. 2019. https://www.researchgate.net/publication/336252989_Impact_of_Climate_Change_on_Productivity_of_Orchids.

Dieringer, Gregg. "The Pollination Ecology of *Orchis spectabilis* L. (Orchidaceae)." *Ohio Journal of Science* 82, no. 5 (December 1982): 218–25. https://www.researchgate.net/publication/260416871.

Editors of Time-Life Books. *The Mystical Year*. Mysteries of the Unknown. Alexandria, VA: Time-Life Books, 1992.

Enger, Eldon D., and Bradley F. Smith. *Environmental Science - A Study of Interrelationships*, 8th ed. New York: McGraw-Hill, 2002.

Fay, Michael F. "Orchid Conservation: How Can We Meet the Challenges in the Twenty-First Century?" *Botanical Studies*, no. 59 (June 5, 2018), article no. 16. https://doi.org/10.1186/s40529-018-0232-z.

Forrester, Nicole. "Inside the Trap of a Yellow Lady's Slipper Orchid (*Cypripedium parviflorum* var. *pubescens*): The Effects of 'Light Windows' and Flower Orientation on the Behavior of a Native Bee (*Andrena Macra*)." Paper 421. Undergraduate honors thesis, College of William & Mary, 2011. https://scholarworks.wm.edu/honorstheses/421.

Hargreaves, Dorothy, and Bob Hargreaves. *Tropical Blossoms of the Pacific*, vol 1. Lahaina, HI: Ross-Hargreaves, 1970.

Hayden, John W. "Wildflower of the Year 2016 Downy Rattlesnake Plantain (*Goodyera Pubescens*)." *Virginia Native Plant Society* (blog), accessed March 10, 2023. https://vnps.org/wildflower-of-the-year-goodyera-pubescens-downy-rattlesnake-plantain.

Hemmerly, Thomas E. *Wildflowers of the Central South*. Nashville TN: Vanderbilt University Press, 1990.

Hughes, Nicole M., Giana M. Gigantino, Mary H. Grace, Kevin M. Hoffman, Mary Ann Lila, Brooke N. Willans, and Andrew J. Wommack. "Photosynthetic Profiles of Green, Purple, and Spotted-Leaf Morphotypes of *Tipularia discolor* (Orchidaceae)." *Southeastern Naturalist* 18, no. 4 (Dec. 2019): 641–58. https://doi.org/10.1656/058.018.0415.

"Ice Cream Threatens Turkey's Flowers." BBC World News, August 5, 2003. http://news.bbc.co.uk/2/hi/science/nature/3126047.stm.

Jersáková, Jana, Steven D. Johnson, and Pavel Kindlmann. "Mechanisms and Evolution of Deceptive Pollination in Orchids." *Biological Reviews* 81, no. 2 (2006): 219–35. https://doi.org/10.1017/S1464793105006986.

Kallunki, Jacquelyn A. "*Goodyera*," Flora of North America, last edited Nov. 5, 2020. http://floranorthamerica.org/Goodyera.

"Keys and Fact Sheets: Ceratina Bees." BioNET-EAFRINET, accessed Feb. 29, 2024. https://keys.lucidcentral.org/keys/v3/eafrinet/bee_genera/key/african_bee_genera/Media/Html_eafrica/Ceratina_bees.htm.

Lawler, L. J. "Useful Orchids." *American Orchid Society Bulletin* 54, no. 6 (1985): 690–94.

Lloyd, D. G. "Variation Strategies of Plants in Heterogenous Environments." *Biological Journal of the Linnean Society* 21, no. 4 (1984): 357–85. https://doi.org/10.1111/j.1095-8312.1984.tb01600.x.

Longhena, Maria. *Ancient Mexico: The History and Culture of the Maya, Aztecs, and Other Pre-Columbian Peoples*. Vercelli, Italy: Stewart, Tabori & Chang, 1998.

Magrath, Lawrence K. "*Liparis liliifolia*." Flora of North America, last edited Nov. 5, 2020. http://floranorthamerica.org/Liparis_liliifolia.

Martin, Fredericka. *Frances Toor's New Guide to Mexico including Lower California*, 8th ed. New York: Crown Publishers, 1967.

Masonick, Paul, Amy Michael, Sarah Frankenberg, Wolfgang Rabitsch, and Christiane Weirauch. "Molecular Phylogenetics and Biogeography of the Ambush Bugs (Hemiptera: Reduviidae: Phymatinae)." *Molecular Phylogenetics and Evolution*, no. 114 (Sept. 2017): 225–33. https://doi.org/10.1016/j.ympev.2017.06.010.

Mattrick, Christopher. "*Liparis liliifolia* (L.) L. C. Rich. Ex Lindley Lily-Leaved Twayblade: Conservation and Research Plan for New England." Framingham, MA: New England Plant Conservation Program, 2004. https://www.nativeplanttrust.org/documents/75/Liparis_liliifolia.pdf.

McCormick, Melissa, Dennis Whigham, and John P. O'Neill. "Mycorrhizal Diversity in Photosynthetic Terrestrial Orchids." *New Phytologist* 163 (June 7, 2004): 425–38. https://doi.org/10.1111/j.1469–8137.2004.01114.x.

Mergen, Daryl E. "*Cypripedium parviflorum* Salisb. (Lesser Yellow Lady's Slipper): A Technical Conservation Assessment." Lakewood, CO: USDA Forest Service, Rocky Mountain Region, 2006. https://www.fs.usda.gov/Internet/FSE_DOCUMENTS/stelprdb5206980.pdf.

Miura, Chihiro, Miharu Saisho, Takahiro Yagame, Masahide Yamato, and Hironori Kaminaka. "*Bletilla Striata* (Orchidaceae) Seed Coat Restricts the Invasion of Fungal Hyphae at the Initial Stage of Fungal Colonization." *Plants* 8, no. 8 (August 11, 2019): article # 280, 1–11. https://doi.org/10.3390/plants8080280.

NatureServe Explorer. s.v. "*Cypripedium parviflorum* var. *pubescens*: Large Yellow Lady's-slipper." page last published Feb. 2, 2024. https://explorer.natureserve.org/Taxon/ELEMENT_GLOBAL.2.134081/Cypripedium_parviflorum_var_pubescens.

Nuttall, George H. F. "Symbiosis in Animals and Plants." *American Naturalist* 57, no. 652 (1923): 449–75. https://www.jstor.org/stable/2456396.

Nyffeler, Martin, and Georg Benz. "Spiders in Natural Pest Control: A Review." *Journal of Applied Entomology* 103, no. 1–5 (1987): 321–39. https://doi.org/10.3929/ethz-a-005802284.

Odum, Eugene P., and Gary W. Barrett. *Fundamentals of Ecology*, 5th ed. Belmont, CA: Thomson Brooks/Cole, 2005.

Ossenbach, Carlos. "History of Orchids in Central America Part I: From Prehispanic Times to the Independence of the New Republics." *Harvard Papers in Botany* 10, no. 2 (January 2009): 183–226. https://doi.org/10.3100/1043-4534(2005)10[183:HOOICA]2.0.CO;2.

Pace, Matthew, and Kenneth Cameron. "The Systematics of the *Spiranthes cernua* Species Complex (Orchidaceae): Untangling the Gordian Knot." *Systematic Botany* 42, no. 4 (Dec. 2017): 640–69. https://doi.org/10.1600/036364417X696537.

Pemberton, Robert W. "Pollination of Slipper Orchids (*Cypripedioideae*): A Review." *Lankesteriana* 13, no. 1–2 (Aug. 2013): 65–73. https://doi.org/10.15517/lank.v0i0.11539.

Pijl, L. van der, and Calaway H. Dodson. *Orchid Flowers: Their Pollination and Evolution*. Coral Gables, FL: Fairchild Tropical Gardens and the University of Miami Press, 1966.

Plant World, The. World Book Encyclopedia of Science, vol. 5. Chicago, IL: World Book, 1992.

Rose, Sarah. *Spiders of North America*. Princeton Field Guides. Princeton, NJ: Princeton University Press, 2022.

Sansom, William. *A Book of Christmas*. New York: McGraw-Hill, 1968.

Schmidt, John M., and Ann E. Antlfinger. "The Level of Agamospermy in a Nebraska Population of *Spiranthes cernua* (Orchidaceae)." *American Journal of Botany* 79, no. 5 (1992): 501–7. https://doi.org/10.1002/j.1537–2197.1992.tb14585.x.

Selosse, Marc-André, Julita Minasiewicz, and Bernard Boullard. "An Annotated Translation of Noël Bernard's 1899 Article 'On the Germination of *Neottia Nidus-Avis*.'" *Mycorrhiza* 27, no. 6 (August 2017): 611–18. https://doi.org/10.1007/s00572–017–0774-z.

Sheviak, Charles J., "*Cypripedium parviflorum* var. *pubescens*," Flora of North America, last edited Dec. 14, 2022. http://floranorthamerica.org/Cypripedium_parviflorum_var._pubescens.

———. "A New *Spiranthes* from the Grasslands of Central North America." *Botanical Museum Leaflets, Harvard University* 23, no. 7 (1973): 285–97. https://doi.org/10.5962/p.168562.

Sheviak, Charles J., and Paul M. Catling. "*Galearis spectabilis*." Flora of North America, last edited Nov. 5, 2020. http://floranorthamerica.org/Galearis_spectabilis.

———. "*Tipularia discolor*." Flora of North America, last edited Nov. 5, 2020. http://floranorthamerica.org/Tipularia_discolor.

Siegal, Carol. "The Hairy Story of Orchid Trichomes." *Orchid Digest* 86, no.1 (Jan.–Mar. 2022): 12–29.

"Speed-Dating the Spiders II: Ghost Spiders." University of Wisconsin-Milwaukee, College of Letters & Science Field Station, *The Buglady* (blog), April 24, 2018. https://uwm.edu/field-station/speed-dating-the-spiders-ii-ghost-spiders.

"*Spiranthes lacera* var. *gracilis*." North Carolina State Extension Gardener Plant Toolbox, accessed Feb. 5, 2024. https://plants.ces.ncsu.edu/plants/spiranthes-lacera-var-gracilis.

Stoutamire, W. A. "Pollination of *Tipularia discolor*, an Orchid with Modified Symmetry." *American Orchid Society Bulletin* 47, no. 2 (1978): 413–15.

Suetsugu, Kenji, Atsushi Kawakita, and Makoto Kato. "Avian Seed Dispersal in a Mycoheterotrophic Orchid *Cyrtosia septentrionalis*." *Nature Plants*, no. 1 (May 5, 2015): article #15052. https://doi.org/10.1038/nplants.2015.52.

Tissue, David, John Skillman, Evan McDonald, and Boyd Strain. "Photosynthesis and Carbon Allocation in *Tipularia discolor* (Orchidaceae), a Wintergreen Understory Herb." *American Journal of Botany* 82, no. 10 (Oct. 1995): 1,249–56. https://doi.org/10.2307/2446247.

US EPA. "Climate Change Indicators in the United States." US EPA, Reports and Assessments, November 6, 2015. https://www.epa.gov/climate-indicators.

Whigham, Dennis F., and John P. O'Neill. "Dynamics of Flowering and Fruit Production in Two Eastern North American Terrestrial Orchids, *Tipularia discolor* and *Liparis liliifolia*." In *Population Ecology of Terrestrial Orchids*, edited by T. C. E. Wells and J. H. Willems, 89–101. The Hague, The Netherlands: SPB Academic Publishers, 1991. Available through Smithsonian Research Online, http://repository.si.edu/xmlui/handle/10088/18419.

Wigginton, Brooks Eliot. *The Foxfire Book: Hog Dressing; Log Cabin Building; Mountain Crafts and Foods; Planting by the Signs; Snake Lore; Hunting Tales; Faith Healing; Moonshining; and Other Affairs of Plain Living*. Garden City, NY: Anchor Press, Doubleday, 1972.

"William John Swainson – Biography, Facts and Pictures." Famous Scientists, accessed March 11, 2023. https://www.famousscientists.org/william-john-swainson/.

Winterringer, Glen S. *Wild Orchids of Illinois*. Popular Science Series, vol. 6. Springfield: Illinois State Museum, 1967.

Zeldovich, Lina. "Bees' and Orchids' Pseudo-Romance Broken by Climate Change." JSTOR Daily, May 8, 2018. https://daily.jstor.org/bees-and-orchids-pseudo-romance-broken-by-climate-change.